Multiple-Choice Questions for A-Level Chemistry

Second Edition

Derek Stebbens

Westminster School

Butterworths

London — Boston

Durban — Singapore — Sydney — Toronto — Wellington

First published 1980
  Reprinted 1981
Second Edition 1984

© Butterworth & Co. (Publishers)Ltd., 1984

British Library Cataloguing in Publication Data

Stebbens, Derek
Multiple-choice questions for A-level chemistry. — 2nd ed.
1. Chemistry — Examiniations, questions, etc.
I. Title
540'.76   QD42

ISBN 0-408-01455-5

Library of Congress Cataloging in Publication Data

Stebbens, D. M. (Derek M.)
Multiple-choice questions for A-level chemistry.
1. Chemistry - Examinations, questions, etc.
I. Title.
QD42.S83   1984   540'.76   83-18877

ISBN 0-408-01455-5

Typeset by Scribe Design
Printed and Bound by Page Bros. Ltd., Norwich, Norfolk

Preface

During the last decade, multiple-choice questions have become an important part of examinations at A level. At first, attempts were made to keep these questions 'secure' and all papers had to be returned to the Examining Boards. The problems of maintaining complete security, pressure from teachers and growing expertise in maintaining a flow of acceptable, tested questions have all combined to produce a growing openness in this area of examining. An increasing number of Boards allow schools to retain the multiple-choice papers and in some cases it is possible to purchase copies of recent papers.

The production of a multiple-choice examination paper is likely to be a lengthy and expensive operation. It would first involve the gathering together, from a writing team, of a number of items considerably in excess of that required for the final paper. A surplus of items is necessary because of the rejection of many at later stages in the process of compiling the paper. These raw items would be seen by a number of experienced teachers who would prepare the items for pre-testing, frequently completely rewriting them in the process. The items that were still being considered at this point would be made up into a series of multiple-choice papers. Each paper would be tested under examination conditions on a large number of pupils. These pupils would be selected to make up a population as near identical as possible to that expected for the real examination in which the questions are to be used.

Computer analysis of the performance of each item and the pre-test as a whole would give information about such factors as the difficulty and discriminating power of individual items. This information would be used to decide which items could go forward as candidates for inclusion in the final paper. The paper would be built up according to a specification which would be different for each Examining Board. The basic framework would be decided by the total number of questions and the number of different types of item to be used. Working within this framework, there would be scope for variety in the order of the item-type, the range of difficulty of items accepted and the overall difficulty of the paper.

There is also the opportunity for considerable variety from Board to Board in the testing of chemical knowledge and understanding. This will be related to the syllabus on which each examination is based, the importance with which each area of the syllabus is viewed and the weight given to the different abilities which an examination of this type can test.

This book is made up of questions that have successfully passed through a selection process similar to that outlined here. The questions used have appeared in the A-level multiple-choice papers of London chemistry, Nuffield chemistry and the chemistry of the Associated Examinations Board. With great generosity, the Boards responsible for these examinations have made available 930 multiple-choice items from past papers. To give more complete coverage of the syllabus of every Board, it has been necessary to write a small number of untested items. These items make up only 5 per cent of the items in this book. Items have been selected from the 'bank' to compose 15 tests of various areas of A-level chemistry. The tests are intended for use at some time after a topic has been completed. Each test contains sufficient items for a selection to be made of those that are suitable for the course being followed.

This book has two unusual features. The first is the use of genuine questions from past papers which have been pre-tested, analysed, accepted and used. The second unusual feature is the discussion of the answers to these questions. All questions are adequately discussed, some at considerable length. This has produced a book where the space occupied by the answers is comparable with the space occupied by the questions. The aim is to help the student develop in general chemical understanding while, at the same time, gaining in confidence and competence at dealing with questions of this type. This approach should be valuable for students in schools or colleges, those working on their own and those who would benefit from guidance over particular difficulties.

I would like to thank Colin Hayes, N.W. Kent College of Technology for his detailed reading of the typescript and for the abundance of helpful suggestions he has made.

DEREK STEBBENS

Contents

Introduction

Five types of multiple-choice question are used in the tests in this book. Some tests use all five types of question but most use less than this. In every test, the order in which the types appears is the same. The order is:

(1) classification sets
(2) multiple-completion (three responses)
(3) multiple-completion (four responses)
(4) multiple-choice
(5) situation sets

Examples of each type of question will be given to help you if you are not already familiar with the directions for answering the question, the appearance of the question and the steps involved in arriving at the correct answer. The correct answer is always one letter chosen from **A, B, C, D** and **E** and is called the 'key'.

1 Classification sets

A classification set begins with the presentation of five similar headings. There is scope for considerable variety in the headings which might, for example, be five different numbers, five different diagrams of apparatus, five different graphs, five different chemicals or five different reactions. A set of questions is based on these five headings and, in each case, you must decide which **one** of the five headings is the best answer. In the complete set, each heading may appear as the correct answer once, more than once or not at all.

Questions 1–3

The ground-state electronic configurations of five elements, **A** to **E**, are shown below:

A $1s^2 2s^2 2p^6 3s^2 3p^6 3d^5 4s^2$

B $1s^2 2s^2 2p^3 3s^2$

C $1s^2 2s^2 2p^6$

D $1s^2 2s^2 2p^6 3s^2 3p^2$

E $1s^2 2s^1$

Select the electronic configuration of an element which

1. is a noble (or inert) gas

2. is most likely to form an oxide with catalytic properties

3. will have the highest first ionization energy of the elements in the GROUP in which it is found

Question 1

You may not have yet made a detailed study of the electronic configuration of atoms. Perhaps, at the moment, you would write the electronic configuration of a noble gas in the form 2.8, 2.8.8 or 2.8.18.8. The 'outer' eight electrons in these configurations are made up of two s electrons and six p electrons within the same 'shell', written $ns^2 np^6$. This is the situation in **C** where the outer electrons are $2s^2 2p^6$. **C** is the correct key.

You would then go on to answer questions **2** and **3**. The answers are not given here because these questions appear in Test 3. Wait until you have worked through Test 3 before looking up the discussion and solution of these problems.

2 Multiple-completion (three responses)

In this type of question, a statement is made and this is followed by three responses. One or more of these responses may follow correctly from and complete the original statement. When it has been decided which responses are correct, the key is chosen by using the following code:

A if **1, 2** and **3** are correct

B if **1** and **2** only are correct

C if **2** and **3** only are correct

D if **1** only is correct

E if **3** only is correct

Directions summarized for question 4				
A	**B**	**C**	**D**	**E**
1,2,3 correct	1,2 only correct	2,3 only correct	1 only correct	3 only correct

Question 4

Substances which would be expected to give a simple line-emission spectrum similar to that of $H(g)$ include

1 $D(g)$

2 $He^+(g)$

3 $Li^{2+}(g)$

The simple line-emission spectrum of gaseous hydrogen atoms is the result of the only electron outside the nucleus of the atom changing from levels of higher energy to levels of lower energy. Deuterium is an isotope of hydrogen so its atoms also contain only one electron. Gaseous deuterium atoms produce a similar spectrum and 1 is correct. Gaseous helium atoms and gaseous lithium atoms contain more than one electron per atom and give more complex spectra. Singly ionized helium, $He^+(g)$, and doubly ionized lithium, $Li^{2+}(g)$, contain only one electron per atom and would be expected to give line-emission spectra similar to that of $H(g)$. 2 and 3 are also correct.

Having decided that 1, 2 and 3 are correct, you should refer to the code. Do not worry about the code before this point. If you can mark your book, you might find it helpful to make a pencil tick or cross against the number of each response as soon as you have decided whether it is correct. With question 4, you should have ticked 1, 2 and 3. When you refer to the code, you will see that the correct answer, or 'key', is **A**.

When you meet this type of question in the tests, the directions inside the box will be printed before the group of questions using the code. If you want more information, you should refer back to the introduction.

3 Multiple-completion (four responses)

This is a similar type of question to multiple-completion (three responses). A statement is made and this is followed by four responses, any one or more than one of which may follow correctly from and complete the original statement. When it has been decided which responses are correct, the key is chosen by using the following code:

ONE or MORE of the responses given are correct. Decide which of the responses is (are) correct. Then choose

A if only **1, 2** and **3** are correct

B if only **1** and **3** are correct

C if only **2** and **4** are correct

D if only **4** is correct

E if some other response, or combination of responses, of those given is correct

Directions summarized for question 5				
A	**B**	**C**	**D**	**E**
1,2,3 only correct	1,3 only correct	2,4 only correct	4 only correct	Some other response or combination of responses is correct

Question 5

Arguments in favour of including hydrogen in the halogen Group of the Periodic Table are

1 hydrogen and chlorine each form a singly charged anion when they combine with the alkali metals

2 molecules of hydrogen are monatomic as are the molecules of all the halogens

3 hydrogen and chlorine can each form covalent compounds with many non-metals

4 hydrogen and chlorine each exist in isotopic forms

Hydrogen and chlorine do form compounds of the type Na^+H^- and Na^+Cl^- when they combine with the alkali metals. The formation of the X^- ion is characteristic of the elements of Group 7 and this is a reason for including hydrogen in the halogen Group. **1** is both a correct statement and a correct reason.

2 is an incorrect statement. Molecules of hydrogen and the halogens are all diatomic: H_2, F_2, Cl_2, Br_2 and I_2.

3 is a correct statement. Hydrogen and chlorine do both form covalent compounds with many non-metals. But so do oxygen, sulphur, carbon, nitrogen and phosphorus, for example, and they are not in Group 7. **3** is not a correct reason for including hydrogen in the halogen Group. If the question had begun 'Arguments in favour of including hydrogen in Group 7 rather than in Group 1 of the Periodic Table are' then **3** would be a correct reason.

4 is a correct statement but most elements can exist in isotopic forms. This is not a reason for including hydrogen in the halogen Group.

This is the point where you refer to the code. Only **1** is a correct reason for including hydrogen in the halogen Group. If you had been putting pencil ticks and crosses against the number of each response, the position should be $\checkmark \times \times \times$. When you refer to the code, you will see that the correct key is **E**.

When you meet this type of question in the tests, the directions inside the box will be printed before the group of questions using this code. If you want more information, you should refer back to this part of the introduction.

4 Multiple-choice

In a question of this type, a statement is made, or a question is asked, and this is followed by five responses. You are required to select the **one** that is most correct, most complete or follows most closely from the original statement. You are choosing the **best** response of those given.

Question 6

The table shows the first four ionization energies (in kJ mol^{-1}) for five elements P, Q, R, S and T (the letters are not chemical symbols).

Ionization energy

	1st	*2nd*	*3rd*	*4th*
P	1090	2400	4600	6200
Q	500	4600	6900	9500
R	740	1500	7700	10 500
S	800	2400	3700	25 000
T	580	1800	2700	11 600

Which of the following pairs is likely to be in the same group of the Periodic Table?

A P and Q

B Q and R

C R and T

D R and S

E S and T

Successive ionization energies give information about the number of 'outer' electrons possessed by atoms of an element. The increases are relatively small until all the electrons in a 'shell' have been removed. There is then a relatively large increase in ionization energy as the next electron is removed from an inner shell.

P is in Group 4 or beyond as there is no sign of a relatively large increase within the values given. Q is in Group 1 as there is a relatively large increase in ionization energy after the first electron has been removed. The value increases by over nine times. R is in Group 2 as there is a relatively large increase in ionization energy after the second electron has been removed. S and T are both in Group 3 because the relatively large increase in ionization energy takes place after the third electron has been removed.

The pair of elements in the same Group is S and T. The key is **E**.

5   Situation sets

A situation set consists of two or more multiple-choice questions (type 4) based on an experiment, an industrial process, a table of information, a piece of apparatus or some other common theme. Each question contains five responses and you choose the best one of those given as described for question 6 earlier.

Test 1

The Periodic Table

Questions 1–5 concern the following outline of the Periodic Table in which the letters **A** to **E** are not the chemical symbols for the elements concerned.

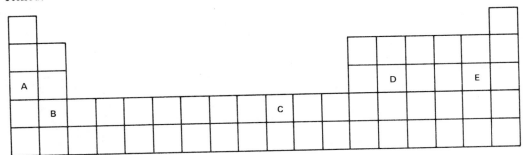

Select, from **A** to **E**, the element which

1. forms coloured ions

2. is a non-metal with a high melting point and high boiling point

3. forms a simple anion in which the element has one negative charge

4. forms a chloride which is hydrolysed by water

5. does NOT react with oxygen even when heated

Directions summarized for questions 6 to 9

A	B	C	D	E
1,2,3 correct	1,2 only correct	2,3 only correct	1 only correct	3 only correct

6. Compared with other metallic elements, the elements of Group 1 have high values for

   1 atomic (molar) volumes

   2 energies required to vaporize one mole

   3 boiling points

7. Hydrides which, when added to pure water, increase the pH include

   1 NH_3

   2 NaH

   3 HBr

8. Methods for preparing anhydrous aluminium chloride include

   1 adding excess aluminium to dilute hydrochloric acid followed by evaporation

   2 adding dilute hydrochloric acid to aluminium hydroxide and evaporating the solution

   3 heating aluminium in a stream of dry chlorine

9. The oxides of non-metallic elements are usually

   1 gaseous at room temperature

   2 covalent in structure

   3 classed as acidic

10. Which of the following compounds give hydrogen chloride on treatment with cold water?

    1 $MgCl_2$

    2 BCl_3

    3 CCl_4

    4 PCl_5

11. Which of the following anhydrous chlorides can be obtained by the action of chlorine on the element?

    1 NaCl

    2 $FeCl_2$

    3 PCl_3

    4 CCl_4

12. 0·01 mole of a chloride, X, was dissolved in water and found to react completely with 20 cm^3 of M silver nitrate solution. X could be

    1 NaCl

    2 $MgCl_2$

    3 $AlCl_3$

    4 S_2Cl_2

13. Along which of these series do the melting points of the elements rise?

    1 He, Li, Be

    2 Be, B, C

    3 Na, Mg, Al

    4 Si, P, S

Directions summarized for questions 10 to 13

A	B	C	D	E
1,2,3 only correct	1,3 only correct	2,4 only correct	4 only correct	Some other response or combination of responses is correct

Directions for questions 14 to 19. Each of the questions or incomplete statements in this section is followed by five suggested answers. Select the best answer in each case.

14. If X represents the element of atomic number 9 and Y the element of atomic number 20, the compound formed between these elements would be

    A covalent, YX

    B ionic, YX

    C covalent, YX_2

    D ionic, YX_2

    E covalent, Y_2X

15. 0·1 mole of each of the following is added separately to one litre of water. Which would produce the solution of lowest pH?

    A Na_2O

    B CaO

    C P_2O_5

    D SO_3

    E SO_2

16. Which of the five chlorides described below, all of the formula type XCl_2, is likely to be the chloride of a Group 2 element?

    A White solid, m.p. 280 °C, b.p. 304 °C, fairly soluble in water to give a colourless neutral solution with a very poor electrical conductivity

    B White solid, m.p. 815 °C, readily soluble in water to give a green-blue solution with good electrical conductivity

    C Red liquid, b.p. 59 °C, insoluble in water and slowly decomposed in contact with it

    D White solid, m.p. 875 °C, readily soluble in water to give a colourless neutral solution with good electrical conductivity

    E White solid, m.p. 672 °C, gives pale green solution in water with good electrical conductivity; the solution darkens on exposure to air

17. In the light of the usual acid–base properties of oxides, which of the following equations suggests that the oxide which appears first on the left hand side of the equation has amphoteric properties?

    A $Li_2O(s) + H_2O(l) \rightarrow 2Li^+(aq) + 2OH^-(aq)$

    B $ZnO(s) + 2OH^-(aq) \rightarrow ZnO_2^{2-}(aq) + H_2O(l)$

    C $CuO(s) + 2H^+(aq) \rightarrow Cu^{2+}(aq) + H_2O(l)$

    D $SO_2(g) + H_2O(l) \rightarrow H^+(aq) + HSO_3^-(aq)$

    E $Cl_2O(g) + 2OH^-(aq) \rightarrow 2ClO^-(aq) + H_2O(l)$

18. In an experiment to determine the formula of a non-metallic bromide of known relative molecular mass, 0·1 mole of the bromide was dissolved in 500 cm^3 of water. 50 cm^3 of this solution reacted exactly with 300 cm^3 of 0·1M $AgNO_3$(aq). If the other element present is denoted by the letter Z, the most probable formula for the bromide is

    A Z_3Br

    B Z_2Br_6

    C ZBr

    D ZBr_3

    E ZBr_6

19. A student prepared a sample of silicon chloride by passing chlorine over heated silicon and collecting the condensed silicon chloride in a small specimen tube. He analysed the chloride by dissolving a known mass of it in water, and titrating the solution with standard silver nitrate solution. The formula of the silicon chloride as obtained by this method was $SiCl_{2.6}$ as against a 'true' formula of $SiCl_4$. Which of the following possible errors could have resulted in this wrong formula?

   A The silicon chloride contained excess, dissolved chlorine.

   B The 'standard' silver nitrate solution was less concentrated than was stated on the label.

   C More silicon chloride than the student supposed was actually used owing to inaccurate weighing.

   D The small specimen tube was not dry.

   E The reaction between the silicon and the chlorine stopped prematurely, leaving some unreacted silicon in the reaction tube.

20. When 0.1 mole of atoms of an element react with chlorine, there is an increase in mass of 7.1 g. The chloride formed has a high melting point. (Relative atomic mass: $Cl = 35.5$)

   The element could be

   A carbon

   B magnesium

   C sodium

   D silicon

   E sulphur

21. An element X forms a liquid chloride which is readily hydrolysed by water. Select from the following the element most likely to be X.

   A Aluminium   D Lithium

   B Beryllium   E Sulphur

   C Carbon

22. An element X, which conducts electricity, forms a chloride XCl_4, which is a colourless volatile liquid. It also forms a solid chloride XCl_2. The element Y, three places above X in the same group of the Periodic Table (i.e., with lower atomic number), does not form a chloride YCl_2. X is probably in Group

   A 3   D 6

   B 4   E 7

   C 5

Questions 23−25 concern the following experiment.

The apparatus shown was set up in the open laboratory (no fume cupboard being available) with the aim of synthesizing the anhydrous chlorides of a number of elements and collecting them in the flask U. In the diagram T is the element to be converted to chloride (in a boat, or in some suitable form, e.g., foil) and U is a two-necked flask (100 cm^3) in which to collect the chloride.

The table below shows the melting and boiling points of some chlorides.

Compound	NaCl	$MgCl_2$	$AlCl_3$	$SiCl_4$
M.p./°C	801	708	sublimes	−70
B.p./°C	1413	1412	178	58

Compound	PCl_3	PCl_5	S_2Cl_2	$FeCl_3$
M.p./°C	−91	sublimes	−80	sublimes
B.p./°C	76	162	136	200

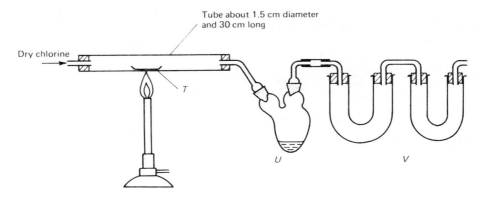

Tube about 1.5 cm diameter
and 30 cm long

Dry chlorine

T

U

V

23. With what would the tubes V be filled?

   A Granular soda-lime

   B Concentrated sulphuric acid

   C Phosphorus pentoxide [phosphorus(V) oxide]
   on glass wool

   D Granular calcium chloride

   E Concentrated sodium hydroxide solution

24. At least one of the following statements is true
about the bunsen burner shown.

   I   It should be operated with the air hole
       closed.

   II It is necessary because all the reactions are
       endothermic.

   III It is usually necessary to provide energy at
       the start of the reaction.

   IV It might cause the joints at the ends of the
       combustion tube to char.

   V   It would minimize the collection, in the
       combustion tube containing T, of any of the
       listed non-metal chlorides.

Which of the above statements are correct?

   A I only

   B II, III, IV and V

   C II, III and IV

   D II, III and V

   E III and V

25. In order to obtain chlorides uncontaminated
with oxides in the flask U, it might be advan-
tageous to displace air from the apparatus
with carbon dioxide before passing chlorine
through. This would be especially important
if the element T were

   A magnesium

   B silicon

   C iron

   D phosphorus

   E aluminium

Test 1   Answers

1. In this form of the Periodic Table, the transition elements come between Groups 2 and 3 of the 'normal' or non-transition elements. Formation of coloured ions is a property common to most transition elements so **C** is the correct choice to answer the first question. Had **C** been two spaces to the right, would it still have been a correct choice?

2. You are probably familiar with the 'diagonal' or 'stair-case' which runs through the Periodic Table and separates metals from non-metals. The diagonal passes to the left of **D**. **D** and **E** are non-metals.

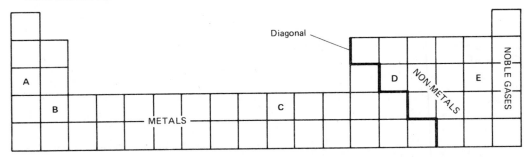

The general trend in structure on going across the Periodic Table from left to right is: Giant structure of atoms (metallic) → giant structure of atoms (non-metallic) → small molecules (non-metallic). Non-metallic elements close to the diagonal tend to have a giant structure of atoms and this gives them high melting and boiling points. **D** is such an element and is the correct choice.

You may well have answered this question by realizing that **D** is silicon. Check in a data book the melting and boiling points of silicon, boron and carbon and locate the Periodic Table positions of boron and carbon. These three elements show most clearly the break between metallic and small molecule structures.

3. The elements of Groups 1 and 2 form only positive ions. Formation of negative ions is a non-metallic property. Some metallic elements close to the diagonal, such as aluminium, form both positive and negative ions and are said to be 'amphoteric'. Many transition elements form positive ions (metallic property) in their low valence (oxidation) states and negative ions (non-metallic property) in their high valence states. Manganese and chromium are examples with ions including Mn^{2+}, Mn^{3+}, Cr^{2+}, Cr^{3+}, MnO_4^{2-}, MnO_4^-, CrO_4^{2-} and $Cr_2O_7^{2-}$. Formation of *simple* negative ions is not a characteristic property of transition elements.

The principal ions of elements to the right of the diagonal carry negative charges. Many of these ions contain a second element, frequently oxygen, and are therefore not regarded as 'simple'. Examples of such ions are SiO_3^{2-}, PO_4^{3-}, SO_3^{2-}, SO_4^{2-}, ClO^-, ClO_3^- and ClO_4^-. How many of these ions can you name?
The *simple* negative ions formed by non-metallic elements show a clear pattern in the value of the charge they carry. Group 7 elements form ions of charge -1, (Cl^-, Br^-, I^-). Group 6

elements form ions of charge $-2(O^{2-}, S^{2-})$.
Group 5 elements form ions of charge $-3 (N^{3-})$.
E is a Group 7 element and is the correct choice
to answer the question.

4. Chlorides of the most metallic elements, such
as **A** and **B** in this problem, produce neutral
solutions in water. The giant structure of ions
in these solid chlorides become aqueous solutions of ions without hydrolysis.

   Chlorides of less metallic elements — those
nearer the diagonal and transition elements —
are frequently extensively hydrolysed by water.
In this question, however, you should look
first to see if **D** or **E** fit the description. **D** and
E are non-metals and chlorides of such elements
are usually completely hydrolysed by water. **D**
is silicon and the chloride of this element is
rapidly hydrolysed by water in a very vigorous,
exothermic reaction:

$$SiCl_4 + 2H_2O \rightarrow SiO_2 + 4HCl$$

E is chlorine and the molecule Cl_2 could be
regarded as 'chlorine chloride'. Chlorine water
contains chlorine which is partly in the form of
hydrated chlorine molecules, these being
responsible for the pale green colour. Some of
the chlorine molecules have been hydrolysed
to hydrochloric acid and hypochlorous acid
[which you may call chloric(I) acid on your
course] :

$$ClCl + H_2O \rightarrow HCl + HOCl$$

Although **E** is a possible answer to this question,
D is the best answer as there is no doubt that
$SiCl_4$ would be regarded by all chemists as a
'chloride'.

5. Oxygen, like other non-metals, can form compounds with elements from all parts of the
Periodic Table. Oxygen is one of the most reactive of non-metals and oxides can usually be
formed by direct combination.

   Group 1 elements, such as **A**, combine with
oxygen from the air at room temperature —
freshly cut sodium viewed under a binocular
microscope is a memorable sight. Group 2
elements, such as **B**, require some persuasion to
react rapidly in oxygen as the initial coating of
oxide slows down further attack. Once started,
however, they burn without further assistance.

Some less metallic elements, such as aluminium, tin, lead and the transition elements (**C**),
are only oxidized on the surface and even this
requires strong heating in air or oxygen. If the
temperature is sufficiently high, many of these
elements will burn in oxygen. The non-metal
silicon (**D**) behaves in a similar way.

There are a few elements which cannot be
made to react with oxygen at any temperature.
The oxides of these elements have to be made
by other methods. Chlorine (**E**) is such an element and is the correct choice in this question.
Can you find any more elements whose oxides
have to be made by methods other than direct
combination? Are there some elements which
do not form oxides at all?

6. The Group 1 elements have high molar volumes
compared with other metallic elements. There
are several reasons for this including:

   (1) The atoms of alkali metals are relatively
large. If you look at a graph of atomic
radius (or metallic radius) against atomic
number you will see this is a 'Periodic'
property with a high value for each Group
1 element, followed by a steady decrease
with increasing atomic number. Another
high value appears at the next Group 1
element and the pattern repeats at periodic
intervals.
   (2) The atoms of Group 1 elements are not
packed as closely as the atoms of many
other metallic elements. The atoms of
Group 1 elements pack in the body-
centred cubic arrangement in which each
atom has only eight nearest neighbours.
The two other simple structures which
metallic elements might take up are closer
packed with twelve nearest neighbours in
both cases.
   (3) The metallic bonding in Group 1 elements
is weak relative to that in other metallic
elements. Metallic bonding uses the 'outer'
electrons of which there is only one per
atom for Group 1 elements.

Statements 2 and 3 are concerned with the
complete break up of the metallic structure
which occurs on vaporization. The weak
metallic bonding and poor packing in Group 1
elements makes it relatively easy to vaporize

them and the quantities mentioned in **2** and **3** have low values.

Only **1** is correct so **D** is the key.

7. Both ammonia and sodium hydride increase the concentration of hydroxide ions by reaction with water. This is equivalent to an increase in pH. Ammonia forms the weakly ionized ammonium hydroxide while sodium hydride forms fully ionized sodium hydroxide. Possible equations are:

$$NH_3(g) + H_2O(l) \rightleftharpoons NH_4^+(aq) + OH^-(aq)$$

$$Na^+H^-(s) + H_2O(l) \rightarrow$$
$$Na^+(aq) + OH^-(aq) + H_2(g)$$

Hydrogen bromide reacts with water to form hydroxonium (hydronium, oxonium) ions, the hydrated hydrogen ion. This is equivalent to a decrease in pH:

$$HBr(g) + H_2O(l) \rightarrow H_3O^+(aq) + Br^-(aq)$$

The correct key is **B**.

8. Aluminium chloride is extensively hydrolysed by water, a possible equation being

$$AlCl_3 + 3H_2O \rightleftharpoons Al(OH)_3 + 3H^+ + 3Cl^-$$

This reaction makes it impossible to obtain $AlCl_3$ by any process involving evaporation from aqueous solution. Evaporation drives off HCl gas as well as water vapour and increases the extent of hydrolysis until all that remains is solid aluminium hydroxide or oxide.

Methods **1** and **2** would both produce aqueous $AlCl_3$ at their first stages but are not methods for obtaining the anhydrous solid. Method **3** would give the anhydrous solid so the correct key is **E**.

$$Al(s) + 1\tfrac{1}{2}Cl_2(g) \rightarrow AlCl_3(s)$$

Look up the preparation of $AlCl_3$ if you do not know it already and decide if the state symbol (s) given in the above equation is correct in both cases.

Anhydrous $AlCl_3$ can be obtained by heating aluminium in a gas other than chlorine. What is this other gas and can you write an equation for the reaction?

9. The bonding in compounds between non-metals is covalent. The oxides of non-metals are examples of this type so statement **2** is correct.

The fact that a compound is covalent in bonding does not mean that it will be gaseous at room temperature. To be gaseous, a substance must be covalent and possess a relatively small, non-polar molecule. Some oxides, such as SiO_2, possess a giant covalent structure while others, such as P_4O_{10} or H_2O are relatively large or polar. These oxides are solids or liquids. Statement **1** is incorrect.

Some oxides of non-metals dissolve in water to form acidic solutions. SO_2 dissolving to form sulphurous acid, H_2SO_3, or CO_2 dissolving to form carbonic acid, H_2CO_3, are examples. It could not be said that non-metal oxides 'usually' react in this way with water but it would be correct to say that they usually react with a strongly ionized base such as the oxide or hydroxide of sodium or calcium. This reaction with a strongly ionized base is the usual criterion for classifying an oxide as acidic. A salt is formed in this reaction, the non-metal oxide appearing in the negative ion. The non-metal oxide would be classified as acidic.

2 and **3** are correct so the key is **C**.

The equations below show several ways of representing the behaviour of carbon dioxide when it acts as an acidic oxide, forming the carbonate or hydrogencarbonate ion. Later in your course you will probably study the idea of acids and bases in more detail and may wish to return to these seven equations to see how they fit in with your greater understanding of acid–base reactions.

$$CO_2 + Ca^{2+}O^{2-} \rightarrow Ca^{2+}CO_3^{2-}$$
$$CO_2 + O^{2-} \rightarrow CO_3^{2-}$$
$$CO_2 + Ca(OH)_2 \rightarrow Ca^{2+}CO_3^{2-} + H_2O$$
$$CO_2 + 2OH^- \rightleftharpoons CO_3^{2-} + H_2O$$
$$CO_2 + OH^- \rightleftharpoons HCO_3^-$$
$$CO_2 + 3H_2O \rightleftharpoons CO_3^{2-} + 2H_3O^+$$
$$CO_2 + 2H_2O \rightleftharpoons HCO_3^- + H_3O^+$$

10. This is a question about hydrolysis of chlorides. Hydrolysis occurs readily with non-metal chlorides and with chlorides of metals close to the diagonal separating metals from non-metals (*see* the answer to question 2).

Magnesium is not such a metal and its chloride is not hydrolysed by cold water. It should be noted that magnesium chloride is hydrolysed at high temperature and this makes it impossible to obtain the anhydrous salt by simple evaporation of its aqueous solution.

The other three compounds are non-metal chlorides but the correct key is not **E** as tetrachloromethane is resistant to hydrolysis. This exception to the general rule about hydrolysis of non-metal chlorides makes the correct key **C**.

Note that the question says 'hydrogen chloride'. It could equally well have said 'an acidic solution' or 'hydrochloric acid' or 'hydrogen ions'.

11. The extremely strong bonding in the giant structures of both diamond and graphite make carbon a very unreactive element. Chlorine, one of the most reactive of non-metals, does not react with carbon so **4** is incorrect.

Sodium, iron and phosphorus all react readily with chlorine but this does not make the correct key **A** as chlorine converts iron to $FeCl_3$. The correct key is **B**.

There is a gas which will form $FeCl_2$ when passed over heated iron. What is this gas?

Which of these chlorides is converted to a higher chloride, XCl_5, if excess chlorine is used?

Note that the question says 'can be'. Which of these chlorides occurs naturally and does not need to be made by a chemical reaction?

12. 20 cm^3 of molar silver nitrate contains $20/1000$ or 0.02 mole of silver ions. These react with chloride ions in the ratio of 1 to 1 by moles.

$$Ag^+(aq) + Cl^-(aq) \rightarrow AgCl(s)$$

This means that 0.01 mole of the chloride, X, contains 0.02 mole of chlorine, Cl. 1 mole of the chloride contains 2 moles of chlorine. The two moles of chlorine could already be present as chloride ions, as it is in $MgCl_2$, or could be released by hydrolysis, as it is with S_2Cl_2. 2 and 4 are correct so the key is **C**.

13. This question can be answered correctly if you know the pattern of melting-point change within the Periodic Table. On going across a period, the structure of an element changes from metallic bonding, through giant covalent (with earlier periods), to small molecules, as mentioned in the answer to question 2. The period finishes with the smallest possible molecule, the single atom of a noble gas.

Within a period, the melting point starts fairly low and increases across the period as the number of electrons available for metallic bonding increases. If one or more elements possess a giant covalent structure, the melting point peaks here and then drops to low values with the small molecules that follow.

Looking at each set of elements in the question, the structures and numbers of electrons used in giant structures are

1 small molecule, metallic (1 electron), metallic (2 electrons) so melting point rises;
2 metallic (2 electrons), giant covalent (3 electrons), giant covalent (4 electrons) so melting point rises;
3 metallic (1 electron), metallic (2 electrons), metallic (3 electrons) so melting point rises;
4 giant covalent (4 electrons), small molecule, small molecule so melting point drops after silicon.

1, 2 and 3 show a rise so the key is **A**.

If you check the melting points in a data book you will find that the melting point of aluminium is only just larger than that of magnesium. This is because melting point is not the best indicator of the strength of bonding in a structure. The boiling point or the energy required to vaporize 1 mole (*see* question 6) give a far better indication of the strength of an element's structure. Find gallium in the Periodic Table and decide the type and strength of its structure. Then look up the melting and boiling points of gallium and decide which value gives a better idea of the strength of bonding in this element's structure.

14. You must first decide from which Group in the Periodic Table each element comes. X is in Group 7 and Y is in Group 2. As X is a non-metal and Y is a metal, the bonding between them must be ionic. Only **B** or **D** could be correct. The charges on the ions must be as shown by X^- and Y^{2+} so the formula of the compound will be YX_2 and **D** the correct key.

15. The solution of lowest pH that you are asked to identify will be the one that is most acidic. **A** and **B** produce alkaline solutions so the choice is between **C**, **D** and **E**. The acids formed by P_2O_5 and SO_2 are only weakly ionized and the correct choice is SO_3, key **D**. This is because sulphuric acid [which may be called sulphuric (VI) acid on your course], formed by the action of water on SO_3, is a very strongly ionized acid in its first stage of ionization.

$$H_2SO_4(aq) + H_2O(l) \rightarrow$$
$$H_3O^+(aq) + HSO_4^-(aq)$$
$$[\text{complete ionization}]$$

You might like to refer to a data book to compare how strongly ionized sulphuric acid and phosphoric acid are in aqueous solution. You may find this in an index under 'dissociation constants' or 'equilibrium constants' or 'ionization constants'.

16. A Group 2 chloride will be a colourless solid of ionic structure with the formula $X^{2+}2Cl^-$. It will have a high melting point and a high boiling point and be readily soluble in water. The aqueous solution will contain the ions $X^{2+}(aq)$ and $Cl^-(aq)$ which will give good electrical conductivity. No hydrolysis will take place so the solution will be neutral.

    D mentions some of these expected properties and contains no incorrect statement — it is the answer to this question. Two other chlorides are close to being correct choices but the information given suggests that these are chlorides of transition elements. What are the keys of these two chlorides?

17. An amphoteric oxide is one that shows both acidic and basic properties. Metallic oxides are usually basic and non-metallic oxides are usually acidic.

    A shows Li_2O behaving as expected for an oxide of Group 1 or 2 where the oxide dissolves in water to form an alkaline solution. As only one property is given each time, it is necessary to look for the opposite of expected behaviour and this occurs in **B**. Zinc oxide would be expected to be basic but it is shown here acting as an acidic oxide (*see* the third paragraph of the answer to question 9). Before deciding that **B** is the correct key it is best to look through the remaining equations in case

one has missed some point that the question is testing. **C** has CuO behaving, as expected, as a basic oxide while **D** and **E** have non-metal oxides behaving, as expected, in an acidic way. **B** is the correct key.

18. Using the information given, the number of moles of bromide used in the reaction is 0·01. The non-metal bromide will hydrolyse to produce bromide ions which will react with $AgNO_3(aq)$.

$$Ag^+(aq) + Br^-(aq) \rightarrow AgBr(s)$$

The number of moles of $Ag^+(aq)$ required to react with the bromide ions is 0·1 × 300/1000 or 0·03. As silver and bromide ions react in the ratio 1 mole to 1 mole, 0·01 mole of the bromide of Z produce 0·03 mole of bromide ions by hydrolysis. This would be the case if the formula is ZBr_3. The correct key is **D**.

19. The experimental formula suggests that less chlorine is present than expected. The first three suggestions would all give an answer with more chlorine than expected because more silver nitrate solution would be needed than the 'ideal' value.

    D would give an incorrect formula because some of the $SiCl_4$ would react with the dampness in the specimen tube. HCl gas would be lost during this hydrolysis leaving SiO_2 mixed with the remaining $SiCl_4$. The $SiCl_4$ that was weighed out would be 'diluted' by the oxide that contained no chlorine at all and hence too low a value would be obtained for the quantity of chlorine present in pure silicon chloride. The correct key is **D**.

    Checking **E** just in case, this will not affect the formula as the substance analysed is the $SiCl_4$ that is distilled over from the reaction tube. It does not matter if some silicon remains unreacted.

20. The increase in mass of 7.1 g is due to the chlorine that has been taken up by 0.1 mole of the element on forming the chloride. The number of moles of chlorine atoms is 0.2. This shows that the empirical formula of the chloride is ECl_2.

    Of the five elements between which choice has to be made, carbon does not react directly with chlorine while sodium and silicon do react but form chlorides of formula $NaCl$ and $SiCl_4$. Magnesium chloride has the formula $MgCl_2$ while sulphur forms one chloride of formula SCl_2. As sulphur is a non-metal, its chloride will be composed of small molecules which will give it a low melting point rather than the high melting point mentioned in the question. Magnesium chloride is a compound of a metal and a non-metal so it will be composed of a giant structure of ions. This will give the compound a high melting point and make magnesium, key **B**, the correct choice.

21. Only two of the elements in the list form liquid chlorides. These are carbon and sulphur. The chloride of carbon is resistant to hydrolysis (see the answer to question 10) while sulphur, like most non-metal chlorides, is readily hydrolysed by water. The correct key is **E**.

22. The electrical conductivity of element X indicates that it is a metal. It cannot be graphite as there has to be another element Y that is three places above X in the same Group. X must be an element below the diagonal on the right-hand side of the Periodic Table on page 10. X cannot be an element in the odd-numbered Groups 3, 5 and 7 as chlorides of these elements such as $InCl_3$, PCl_5 and ICl_3, contain an odd number of chlorine atoms in their molecules.

    X must be in Group 4 or Group 6. If X is in Group 6, this would make Y, which is three places above X, either oxygen or sulphur. These two elements do form chlorides OCl_2 and SCl_2. This evidence rules out Group 6.

    This leaves Group 4 where the top elements, carbon and silicon, are non-metals while the lower elements, tin and lead, show many metallic properties. Tin or lead are very suitable choices for element X as the higher chlorides of these elements, $SnCl_4$ and $PbCl_4$, are covalent liquids (non-metallic property) while the lower chlorides, $SnCl_2$ and $PbCl_2$, are solids which conduct electricity, when molten (metallic property). The chlorides of tin and lead have the right formulae and properties for one of these elements to be X. This would make Y either carbon or silicon and the failure of these

higher elements to form chlorides CCl_2 and $SiCl_2$ helps to confirm this choice. The Group involved is 4 and the key is **B**.

23. The purpose of the tubes V is to absorb all the chlorine that remains after passing over the heated elements in T. **B**, **C** and **D** do not absorb chlorine — they are all drying agents. As well as being inappropriate, concentrated sulphuric acid would be dangerous in the apparatus shown as it would be blown out by the gas flow. Concentrated sodium hydroxide solution would absorb chlorine but, being a liquid, would also be blown out of the apparatus.

    Soda lime, which is a solid mixture of sodium hydroxide and calcium hydroxide, would absorb chlorine and would allow gas to flow through as it is granular. So **A** is the correct key. Can you find out how soda lime is made and can you write an equation for one of its components reacting with chlorine?

24. The air hole of the bunsen should be at least partly open to avoid the yellow, smoky flame so **I** is incorrect. Combination with chlorine is an exothermic reaction (**II** incorrect) but energy is usually required to start the combination (**III** correct). Thre is some doubt in the diagram about the distance between the bunsen and the joints at the end of the combustion tube but it is not necessary to decide if **IV** is correct. **V** is certainly correct and the bunsen may well be momentarily moved to the right along the combustion tube in order to vaporize any chloride which settles there.

    Correct statements are **III**, **IV** and **V**, or **III** and **V**. Looking at the choices in the question, only one of these is given so the key must be **E**.

25. The spontaneous combustion of phosphorus in the oxygen of the air makes contamination by phosphorus oxides very likely. This contamination could be reduced by displacing the air with a suitable gas before passing chlorine. The correct key is **D**. You might like to check in a reference book or data book to see how many possible compounds can be made from phosphorus, chlorine and oxygen (you should be able to find five). How many of these are likely to collect in U if air is not displaced from the apparatus?

Test 2

Atomic and molecular masses

Directions summarized for questions 1 to 6

A	B	C	D	E
1,2,3 only correct	1,3 only correct	2,4 only correct	4 only correct	Some other response or combination of responses is correct

1. Which of the following masses of gas would occupy about 3 dm^3 at 25 °C and 1 atmosphere? (1 mole of gas occupies 24 dm^3 at 25 °C and 1 atmosphere; C = 12, O = 16, S = 32, Ar = 40)

   1 8.0 g of sulphur dioxide

   2 5.5 g of carbon dioxide

   3 4 g of oxygen

   4 10 g of argon

2. Which of the following aqueous solutions contains the same number of solute particles as are contained in 250 cm^3 of 2M sodium chloride?

   1 1 dm^3 of M ethanol, C_2H_5OH

   2 250 cm^3 of 3M calcium chloride, $CaCl_2$

   3 500 cm^3 of M hydrochloric acid, HCl

   4 500 cm^3 of M ethanoic acid, CH_3CO_2H

3. If L is the Avogadro constant, correct statements about a 2M solution of copper(II) nitrate include that

   1 1 dm^3 of solution contains a total of $6L$ ions of Cu^{2+} and NO_3^-

   2 1 dm^3 of solution contains $2L$ NO_3^- ions

   3 500 cm^3 of solution is 2M with respect to copper(II) ions

   4 500 cm^3 of solution is 2M with respect to NO_3^- ions

4. On complete oxidation, 1 mole of an organic compound gave 4 moles of water. The compound could be

   1 methanol

   2 propane

   3 ethylene (ethene)

   4 but-1-ene

5. The vapour of an organic compound requires three times its own volume of oxygen for complete combustion, and produces twice its own volume of carbon dioxide. Which of the following compounds would give these results?

1 CH_3CHO

2 C_2H_5OH

3 CH_3CO_2H

4 C_2H_4

6. The rates of diffusion of gases will increase with

1 increase in pressure

2 increase in relative molecular mass

3 increase in temperature

4 addition of platinum as a catalyst

Directions for questions 7 to 18. Each of the questions or incomplete statements in this section is followed by five suggested answers. Select the best answer in each case.

7. In the nineteenth century, there was considerable doubt over the formula of zirconium oxide. The relative atomic mass of zirconium (Zr) was thought to be about 90. The relative molecular mass of the volatile chloride was found to be 236 (Cl = 35.5). This evidence suggests that the most likely formula for the oxide of zirconium is

A ZrO

B ZrO_2

C ZrO_3

D Zr_2O_3

E Zr_2O_5

8. 20 cm^3 of a gaseous element X reacts with excess of an element Y to form 40 cm^3 of a gaseous compound of X and Y, all volumes being measured under the same conditions of temperature and pressure.

From this information, it can be deduced that

A the molecule of X contains at least two atoms of X

B the formula of the compound formed is XY

C equal volumes of gases contain equal numbers of molecules

D molecules of X cannot consist of more than two atoms

E X is less dense than the compound of X and Y

9. The following table gives the relative molar mass and percentage of Q for each of three compounds of the element Q.

	Relative molar mass	Percentage of Q
Compound 1	100	48
Compound 2	40	40
Compound 3	64	50

What is the probable relative atomic mass of Q?

A 16

B 32

C 40

D 48

E 50

10. The currently accepted definition of a mole of any substance is 'That amount of the substance which

    A has a mass equal to the formula weight in grammes'

    B would occupy, if gaseous, 22·414 dm³ at a temperature of 273 K and 1 standard atmosphere pressure'

    C has exactly 6·02 × 10²³ particles of the type specified'

    D has exactly the same number of particles, of the type specified, as there are hydrogen atoms in 1 g of hydrogen'

    E has exactly the same number of particles, of the type specified, as there are carbon atoms in 12 g of carbon mass twelve (^{12}C)'

11. Which of the following statements about the Avogadro constant is true?

    A It is the number of electrons required to deposit one mole of atoms of any metallic element from a solution of one of its salts.

    B It is the number of atoms contained in one mole of atoms of any monatomic element.

    C It is the number of grams of any element which contains 6·02 × 10²³ atoms of that element.

    D It is the number of particles (i.e., ions, atoms, molecules) required to make one gram of the substance under consideration.

    E It can be accurately determined by measuring the area a known amount of oleic acid occupies assuming that the layer of acid is one molecule thick.

12. An isotope of the element polonium, of atomic mass 210, is strongly radioactive and each day one two-hundredth part of it changes into an inactive isotope of lead. Approximately how many atoms of lead are formed in one day from one milligramme (10^{-3} g) of ^{210}Po?

(The Avogadro constant L is approximately 6 × 10²³ mol⁻¹.)

    A 1·5 × 10¹⁶

    B 3 × 10¹⁸

    C 1·23 × 10¹⁹

    D 3 × 10²¹

    E 1·2 × 10²²

13. A sample of gas occupied 400 cm³ at s.t.p. Its volume, in cm³, at 45 °C and 700 mmHg pressure would be

    A $400 \times \dfrac{700}{760} \times \dfrac{273}{318}$

    B $400 \times \dfrac{700}{760} \times \dfrac{298}{318}$

    C $400 \times \dfrac{760}{700} \times \dfrac{298}{318}$

    D $400 \times \dfrac{760}{700} \times \dfrac{318}{298}$

    E $400 \times \dfrac{760}{700} \times \dfrac{318}{273}$

14. One mole of an ideal gas, of relative molecular mass M and density D g dm⁻³, occupies a volume of V litres at a temperature of T K. The gas constant is R in appropriate units and n is the number of moles. Which of the following relationships is correct?

    A $PV = \dfrac{M}{D}RT$

    B $PV = \dfrac{D}{n}RT$

    C $PVD = RT$

    D $P\dfrac{M}{D} = RT$

    E $PV = \dfrac{D}{nM}RT$

15. Consider 1 g of each of the following substances at s.t.p. Which occupies the greatest volume?

(H = 1, C = 12, O = 16, F = 19, Ne = 20, S = 32)

A Ethane

B Fluorine

C Hydrogen sulphide

D Oxygen

E Neon

16. Which of the general shapes of graphs drawn below most closely represent the distribution of the velocities of molecules in a sample of gas?

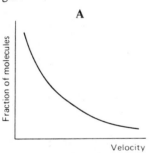

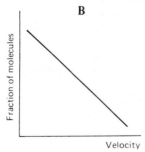

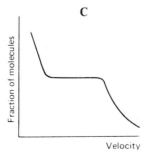

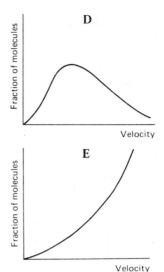

17. Which of the following solutions contains the greatest number of ions?

A 0·8 dm³ of 1·0M $CrCl_3$,$6H_2O$

B 0·6 dm³ of 2·0M $CaCl_2$,$6H_2O$

C 0·8 dm³ of 1·0M $Cr_2(SO_4)_3$,$18H_2O$

D 0·7 dm³ of 0·5M K_2SO_4 ,$Al_2(SO_4)_3$,$24H_2O$

E 0·8 dm³ of 2·0M Na_2SO_4 ,$10H_2O$

18. When heated, a sample of potassium chlorate gave 60 cm³ of oxygen, measured at room temperature and pressure.

$$2KClO_3\,(s) \rightarrow 2KCl(s) + 3O_2\,(g)$$

How many moles of potassium chlorate decomposed?

(The molar volume is 24 dm³ at room temperature and pressure.)

A 0·167 X 10^{-3}

B 0·250 X 10^{-3}

C 1·67 X 10^{-3}

D 2·50 X 10^{-3}

E 3·75 X 10^{-3}

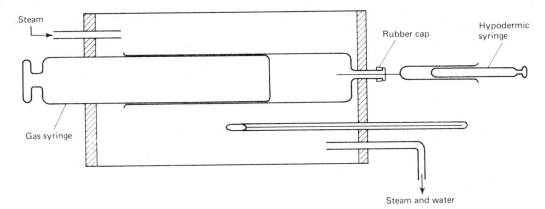

Steam

Rubber cap

Hypodermic syringe

Gas syringe

Steam and water

19. 0.05 g of a gaseous hydrocarbon occupies 80 cm³ at 380 mmHg and 293 K. What is the number of carbon atoms in each molecule of the hydrocarbon? (1 mole of gas molecules occupies about 24 000 cm³ at 760 mmHg and 293 K; C = 12, H = 1)

A 1   D 4

B 2   E 5

C 3

20. What pressure will be developed in a container of volume 20 cm³, containing 1.8 g of water, if it is placed in a furnace at a temperature of 727 °C? (Relative molecular mass: $H_2O = 18$; $R = 0.082$ dm³ atm K^{-1} mol^{-1})

A $1.8 \times \dfrac{0.082}{0.02} \times 10^3$ atm

B $1.8 \times \dfrac{0.082}{0.02} \times 727$ atm

C $\dfrac{0.082}{0.02} \times 10^2$ atm

D $\dfrac{0.082}{0.02} \times 727 \times 10^{-1}$ atm

E $\dfrac{0.082}{0.02} \times 10^{-1}$ atm

Questions 21–23

The apparatus shown above can be used to determine the relative molecular mass of a volatile liquid. The liquid L is drawn into the hypodermic syringe and then introduced into the gas syringe through a self-sealing rubber cap. The liquid turns into vapour, and the volume of vapour produced is recorded. The thermometer reading is also noted.

21. The mass of liquid L introduced into the gas syringe would be found by

A weighing the hypodermic syringe immediately before and after introducing the liquid into the gas syringe

B weighing the hypodermic syringe before filling with liquid and again after introducing the liquid into the gas syringe

C measuring the volume of liquid used and calculating the mass from its density

D measuring the volume of vapour L produced in the gas syringe and calculating the mass using its density

E weighing the gas syringe before and after the experiment

22. The best way of making sure that the vapour of *L* was at the temperature recorded by the thermometer would be to

   A stop the steam supply and allow the gas syringe to cool down to room temperature before reading the volume

   B heat the liquid *L* to a temperature just below its boiling point before drawing it into the hypodermic syringe

   C ensure that the whole apparatus is well lagged with insulating material

   D wait until there was no further change in volume of the gas syringe before recording the volume

   E make sure that the bulb of the thermometer was touching the side of the gas syringe

23. Which of the following liquids could NOT be successfully used in the same way as *L* in this experiment?

Liquid	Relative molecular mass	B.p. /K	Density /g cm^{-3}
A 1,1,1-Trichloroethane	133	347	1·33
B Trichloromethane	120	334	1·48
C Toluene	92	383	0·86
D Tetrachloromethane	154	349	1·59
E Ethyl methyl ketone (butanone)	72	352	0·80

Test 2 Answers

1. Under these conditions, 3 dm³ of any gas would contain 3/24, 1/8 or 0·125 mole of molecules. 1/8 mole of SO_2 is 64/8 = 8 g; 1/8 mole of CO_2 is 44/8 = 5·5g; 1/8 mole of O_2 is 32/8 = 4 g; and 1/8 mole of Ar is 40/8 = 5g. 1, 2 and 3 are all correct so the key is **A**.

2. 250 cm³ of 2M NaCl contains 2 × 250/1000 = 0·5 mole of NaCl. NaCl is fully ionized so the number of moles of particles (ions) present in the solution is 1.
    1 dm³ of M ethanol contains 1 mole of ethanol molecules.
    250 cm³ of 3M $CaCl_2$ contains 3 × 250/1000 moles of $CaCl_2$ which, because of complete ionization, will give 3 × 3 × 250/1000 moles of ions. This is not 1 mole of particles.
    500 cm³ of M HCl contain 500/1000 = 0·5 mole of HCl which, because of complete ionization, will give 1 mole of ions.
    500 cm³ of M ethanoic acid contains 0·5 mole of ethanoic acid. This is only weakly ionized and so the total number of ions and un-ionized molecules will not be as high as 1 mole of particles.
    1 and **3** give 1 mole of particles so the key is **B**.

3. 1 dm³ of 2M $Cu(NO_3)_2$ contains two moles of the compound, which will be fully ionized to give two moles of Cu^{2+} and four moles of NO_3^- ions. Each mole of particles contains L particles so the first statement is correct but the second is not.
    The 500 cm³ in the third and fourth statements is unnecessary. A 2M solution of $Cu(NO_3)_2$ is 2M whatever volume is considered. Because of complete ionization, 2M copper nitrate will be 2M with respect to Cu^{2+} ions but 4M with respect to NO_3^- ions. The third statement is correct but the fourth is not.
    1 and **3** are correct so the key is **B**.

4. There is no need to write equations for the oxidation of the substances in the question. The hydrogen to form the four moles of water must come completely from the organic compound. It is only necessary to find which of the compounds contains eight atoms of hydrogen per molecule. The formulae are CH_3OH, C_3H_8, C_2H_4 and C_4H_8. Only **2** and **4** are correct and the key is **C**.

5. In volumes the reaction is

    1 volume of organic compound + 3 volumes O_2 → 2 volumes CO_2 +

    Using Avogadro's law, volumes can be replaced by number of molecules or number of moles. The reaction becomes

    1 mole of organic compound + 3 moles of O_2 → 2 moles of CO_2 +

    As one mole of organic compound gives two moles of CO_2, each mole of suitable compound must contain two moles of carbon. This is not much help in eliminating possibilities in this question as all four organic compounds contain two carbon atoms per molecule! It is going to be necessary to write an equation for each combustion to find which substances require three moles of O_2 per mole of compound:

$$CH_3CHO + 2·5O_2 \rightarrow 2CO_2 + 2H_2O$$
$$C_2H_5OH + 3O_2 \rightarrow 2CO_2 + 3H_2O$$
$$CH_3CO_2H + 2O_2 \rightarrow 2CO_2 + 2H_2O$$
$$C_2H_4 + 3O_2 \rightarrow 2CO_2 + 2H_2O$$

    The second and fourth equations are correct. The key is **C**.

6. The movement of molecules of a gas through their own molecules, or through molecules of another gas, is called gaseous diffusion. Diffusion is a consequence of the constant, random motion of molecules. The rate of diffusion can be increased by increasing the number of gaseous molecules in unit volume available to move (by an increase in pressure) or by increasing the speed of the molecules (by

an increase in temperature). **1** and **3** are correct, key **B**. [Molecules of higher relative molar mass diffuse more slowly. It is not possible to catalyse the rate of diffusion.]

7. The formula of zirconium chloride will be $Zr_1 Cl_x$ and its relative molecular mass will be $(90 + 35 \cdot 5x)$. As this is also about 226, x must be 4.

   If zirconium retains the same valence (oxidation state) in its oxide, the formula of the oxide will be ZrO_2. This is the 'most likely' formula and the key is **B**.

8. Replacing volumes by number of molecules (or moles) as in question 5 gives the information:

   1 molecule of X →
   $\qquad$ 2 molecules of compound with Y

   The single deduction that can be made from this alone is that one molecule of X must be capable of splitting equally between two molecules of the compound. This is only possible if X contains an even number of atoms. This makes **A** the best of the five alternatives. **D** is certainly incorrect and the other three statements cannot be deduced from the information given.

9. The contribution made by Q to the relative molar mass of each compound is

   compound 1 $\quad 100 \times 48/100 = 48$

   compound 2 $\quad\ \ 40 \times 40/100 = 16$

   compound 3 $\quad\ \ 64 \times 50/100 = 32$

   This is rather a small sample of compounds but it is just possible that there is a compound here that contains only one atom of Q per molecule. This could be compound 2 which would give Q a relative atomic mass of 16. It is necessary to check that this fits with the other two compounds — it does as compound 3 would contain two atoms of Q and compound 1 would contain three atoms of Q.

   The relative atomic mass of Q cannot be more than 16 and it must be a number that divides exactly into 16, 32 and 48. It is likely to be the highest common factor of these numbers, which is 16, but it could, for example, be 8. How many atoms of Q would be present

in a molecule of each compound if this was the case?

Of the alternatives given, only **A** is a possible answer.

10. The current definition of a mole is given in **E**.

11. The only statement that is certain to be correct in every case is **B**. **A** would be correct for metallic elements forming ions of charge +1. **C** and **D** are incorrect. The method mentioned in **E** would give a very approximate value for the Avogadro constant but not an accurate one.

12. The total number of moles of atoms of ^{210}Po at the start of the day is $10^{-3}/210$. At the end of the day, the number of moles of lead atoms formed by decay will be $10^{-3}/(210 \times 200)$. The number of atoms of lead formed is

$$\frac{10^{-3}}{210 \times 200} \times 6 \times 10^{23} = 1 \cdot 43 \times 10^{16}$$

This value is close to that of key **A**.

13. A reduction in pressure from 760 mmHg to 700 mmHg will increase the volume in the ratio 760/700 (Boyle's law). An increase in temperature from 273 K to 318 K will increase the volume in the ratio 318/273. **E** is the answer which shows these changes correctly.

14. The basic form of the ideal gas equation is $PV = nRT$ where n is the number of moles of gas under consideration.

    What is the volume V of n moles of gas of molar mass M g mol^{-1} and density D g dm^{-3}?

$$V = \frac{mass}{density} = \frac{n \text{ mol} \times M \text{ g mol}^{-1}}{D \text{ g dm}^{-3}} = \frac{nM \text{ dm}^3}{D}$$

Substituting for V in the gas equation gives

$$P\frac{nM}{D} = nRT \text{ which is } P\frac{M}{D} = RT$$

This is key **D**.

15. The greatest volume will be occupied by the substance of lowest relative molar mass. The values of the RMM are C_2H_6 (30), F_2 (38), H_2S (34), O_2 (32) and Ne (20). **E** is the correct key.

16. The correct distribution curve should have few molecules with very low velocities, many molecules with velocities in an intermediate range and few molecules with really high velocities. This is represented by graph **D**.

17. The number of moles of water of crystallization is not relevant in this problem. Calculate the number of moles of substance present in the given volume and then scale up by a factor which takes into account the number of moles of ions coming from one mole of the substance if ionization is complete. For example, if $CrCl_3$ is fully ionized into $Cr^{3+}(aq)$ and $3Cl^-(aq)$, 1 mole of $CrCl_3$ will give 4 moles of ions.

    The number of moles of ions in each solution is

    A $0.8 \times 1 \times 4 = 3.2$

    B $0.6 \times 2 \times 3 = 3.6$

    C $0.8 \times 1 \times 5 = 4.0$

    D $0.7 \times 0.5 \times 8 = 2.8$

    E $0.8 \times 2 \times 3 = 4.8$

The correct key is **E**.

18. The number of moles of oxygen, O_2, formed by decomposition is $60/24\,000$. The equation for the reaction shows that this would have come from

$$\frac{2}{3} \times \frac{60}{24\,000} = 1.67 \times 10^{-3} \text{ mole of } KClO_3$$

which is key **C**.

19. If the pressure were doubled to 760 mm Hg, 0.05 g of the hydrocarbon would occupy only 40 cm³. The mass of hydrocarbon that would occupy the molar volume at 760 mm Hg and 293 K is

$$0.05 \text{ g} \times \frac{24\,000 \text{ cm}^3 \text{ mol}^{-1}}{40 \text{ cm}^3} = 30 \text{ g mol}^{-1}$$

The hydrocarbon could not contain as many as 3 carbon atoms per molecule as that would give it a molar mass in excess of 36 g. It could

not contain only one carbon atom per molecule as that would make the formula CH_{18}. There must be two carbon atoms per molecule (key **B**) with a contribution of 24 g to the molar mass. This leaves a contribution of 6 g from hydrogen, showing that the hydrocarbon is ethane, C_2H_6.

20. The gas equation

$$PV = \frac{w}{M} RT$$

can be rearranged in the form

$$P = \frac{wRT}{VM}$$

where $w = 1.8$ g, $R = 0.082$ dm³ atm K^{-1} mol^{-1}, $T = 1000$ K, $V = 0.02$ dm³, $M = 18$ g mol^{-1}

    Substituting these values in the rearranged gas equation gives

$$P = \frac{1.8 \text{ g} \times 0.082 \text{ dm}^3 \text{ atm } K^{-1} \text{ mol}^{-1} \times 1000 \text{ K}}{0.02 \text{ dm}^3 \times 18 \text{ g mol}^{-1}}$$

This can be cancelled down to the expression given in key **C**.

21. The mass of liquid that is introduced by the hypodermic syringe is only about 0.5 g. This small mass is best found by weighings on the hypodermic syringe where the extra mass of the containing syringe, about 10 g, is relatively small. **A** is the correct key.

    B uses masses involving the hypodermic syringe but will not give the mass of liquid injected! **C** would give an approximate value if the syringe is graduated. **D** suggests that the density of the vapour is already known — if this is the case the relative molar mass could be calculated without carrying out the experiment, by using the equation given in question 14. **E** would give the mass of liquid injected if the large syringe could be withdrawn and carefully dried. This is not the best method, however, as the high mass of the large syringe (about 200 g) makes it difficult to measure with sufficient precision the small increase in mass due to the injected liquid.

22. There is a large, rapid increase in volume as soon as the liquid is injected. This is due to the vaporization of the liquid. For some time after this there is a slow increase in volume as the vapour approaches the temperature of the steam jacket. When the volume no longer increases, the vapour is at the temperature of the surrounding steam, this also being the temperature recorded by the thermometer. The correct key is **D**.

23. There is a lot of unwanted information in this question! For the method to be successful, the liquid must be completely converted to vapour which, preferably, is at a temperature at least 20 K above its boiling point so that deviations from the ideal gas laws are small. Looking at the boiling points it can be seen that toluene, with a value above 373 K, could not be used in this experiment. The toluene would not reach its boiling point in this apparatus. The correct key is **C**.

    How would you modify the experiment if you wanted to find the relative molecular mass of toluene by this method?

Test 3

Atomic structure

Questions 1–3

The ground-state electronic configurations of five elements **A** to **E** are shown below:

A $1s^2\,2s^2\,2p^6\,3s^2\,3p^6\,3d^5\,4s^2$

B $1s^2\,2s^2\,2p^6\,3s^2$

C $1s^2\,2s^2\,2p^6$

D $1s^2\,2s^2\,2p^6\,3s^2\,3p^2$

E $1s^2\,2s^1$

Select the electronic configuration of an element which

1. is a noble (or inert) gas

2. is most likely to form an oxide with catalytic properties

3. will have the highest first ionization energy of the elements in the GROUP in which it is found

Directions summarized for questions 4 to 12				
A	**B**	**C**	**D**	**E**
1,2,3 only correct	1,3 only correct	2,4 only correct	4 only correct	Some other response or combination of responses is correct

4. The atomic emission spectrum of hydrogen

    1 shows the frequencies of radiation emitted by excited atoms when they lose energy

    2 consists of several series of lines, each series converging to a limit at the low frequency end

    3 provides information for measuring the ionization energy for the hydrogen atom

    4 provides information about the bond vibrations in the hydrogen molecule

5. The first ionization energy of an element can be found by

    1 measuring the frequency of X-rays produced when it is bombarded with electrons of varying energies

    2 investigating the electrical conduction of its vapour as it is bombarded with electrons of varying energies

    3 studying the diffraction pattern produced when its vapour is bombarded with electrons of high energy

    4 measuring the frequencies of the lines in the most energetic part of its emission spectrum

6. From which of the following, TAKEN TOGETHER, can the ionization energy of hydrogen (in kJ mol^{-1}) be calculated?

    1 The value of Planck's constant (in kJ mol^{-1} s).

    2 The value of the Avogadro constant.

    3 The frequency of the limit of convergence of the lines in the ultraviolet emission spectrum of hydrogen (in s^{-1}).

    4 The number of energy levels in the hydrogen atom.

7. The plot below shows the values of the logarithms of the first five ionization energies [lg(I.E.)] of an element X.

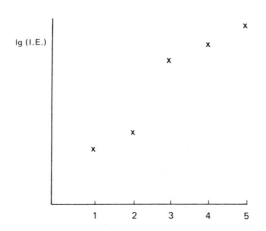

This plot shows

    1 part of the evidence for the arrangement of the electrons of an atom in energy levels or 'shells'

    2 that X cannot have an atomic number less than 12

    3 that X could be an element in Group 2 (an alkaline earth metal)

    4 the rise in ionization energy with increasing average distance of the electron from the nucleus

8. The electronic configuration

could represent ions of

    1 Mn(II)

    2 Fe(III)

    3 Co(III)

    4 Ni(II)

9. Particles having the electronic configuration 2.8.8 include

    1 Ar

    2 Ca^{2+}

    3 Sc^{3+}

    4 Br$^-$

10. Consider the following changes:

    I $M(s) \rightarrow M(g)$

    II $M(s) \rightarrow M^{2+}(g) + 2e^-$

    III $M(g) \rightarrow M^+(g) + e^-$

    IV $M^+(g) \rightarrow M^{2+}(g) + e^-$

    V $M(g) \rightarrow M^{2+}(g) + 2e^-$

The second ionization energy of M could be calculated from the energy values associated with

    1 I + V

    2 V − III

    3 III + IV

    4 II − I − III

11. The unstable nucleus $^{212}_{82}$Pb decays with β particle emission, having a half life of 10 hours. From this it follows that the

1   mass number of the product is 212

2   atomic number of the product is 81

3   fraction of the original isotope remaining after 20 hours is $\frac{1}{4}$

4   nucleus formed is stable

12. Correct statements about the alkali metals include that

1   the first ionization energy decreases with increasing atomic number

2   an unpaired electron is present in an s orbital

3   chemical reactivity increases with increasing atomic number

4   their ions have the electronic configuration of a noble gas

---

Directions for questions 13 to 25. Each of the questions or incomplete statements in this section is followed by five suggested answers. Select the best answer in each case.

---

13. Which of the following is the best description of what Geiger and Marsden found when they tried to detect alpha-particles which they had fired at thin metal foils?

A   All the particles went through the foil whatever metal it was made of.

B   Some of the particles were absorbed by the foil, changing it into a different element.

C   Some of the particles were slightly deflected, the remainder passing through unaffected.

D   Some of the particles bounced back off the foil and the remainder passed through unaffected.

E   Some of the particles went through undeflected and the rest were deflected through a variety of angles, some greater than 90 degrees.

14. The NUCLEUS of $^{23}_{11}$Na contains

A   23 protons and 11 neutrons

B   23 protons and 11 electrons

C   11 protons and 12 neutrons

D   11 protons and 12 electrons

E   12 neutrons and 11 electrons

15. What is the particle X in the following nuclear reaction?

$$^{27}_{13}\text{Al} + ^{1}_{0}\text{n} \rightarrow X + ^{4}_{2}\text{He}$$

A $^{24}_{10}$Ne

B $^{24}_{11}$Na

C $^{24}_{12}$Mg

D $^{24}_{14}$Si

E $^{23}_{11}$Na

16. Which of the following diagrams shows a pattern similar to the emission spectrum of the hydrogen atom in the visible region?

A

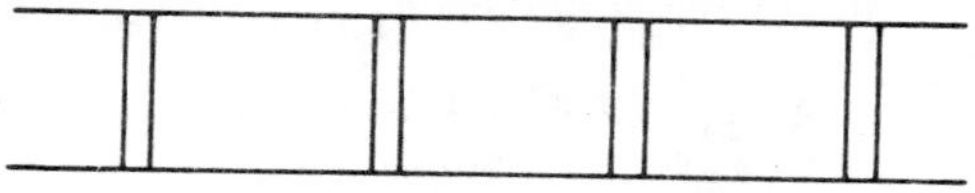

B

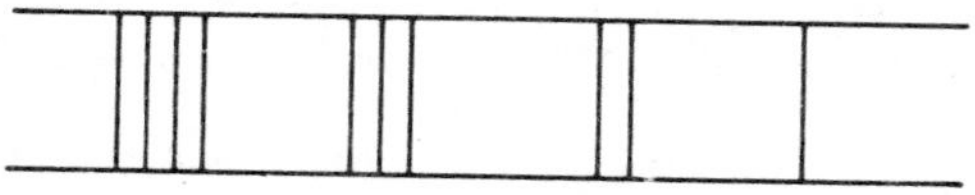

C

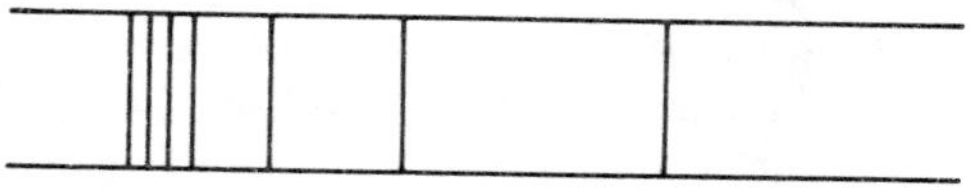

D

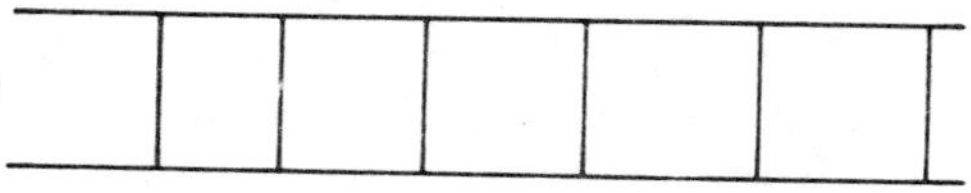

E

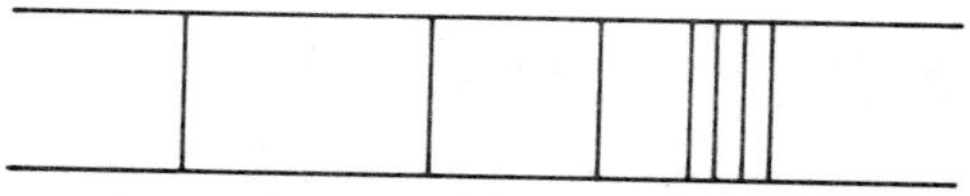

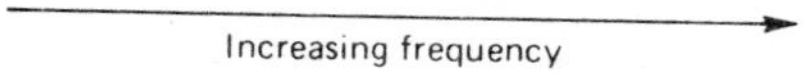

Increasing frequency

17. Which one of the following changes (transitions) between energy (quantum) levels in the hydrogen atom produces the fourth line of the second series in the hydrogen spectrum?

A $n_2 = 5$ to $n_1 = 1$

B $n_2 = 4$ to $n_1 = 1$

C $n_2 = 6$ to $n_1 = 2$

D $n_2 = 5$ to $n_1 = 2$

E $n_2 = 4$ to $n_1 = 2$

18. The atomic number of iron is 26. The electronic structure of the Fe(III) ion may be represented as

A

B

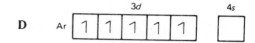

C

D

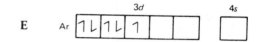

E

19.

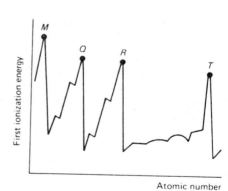

The elements indicated by the letters M, Q, R and T in the diagram above are the

A alkali metals

B alkaline earth metals

C transition elements

D halogen elements

E noble gases

20. The ground-state electronic configurations of five elements are shown below. For which element would you expect the value of the first ionization energy to be the greatest?

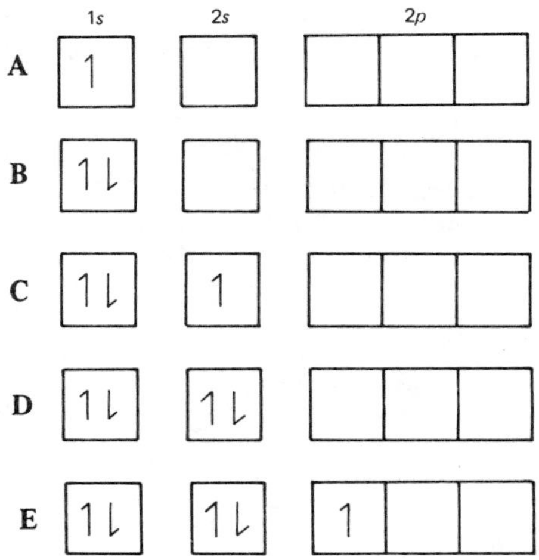

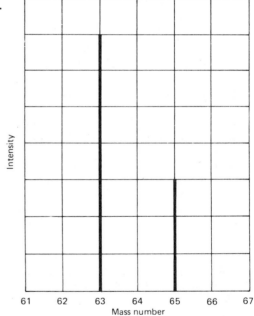

21. Which of the following equations correctly represents the second ionization energy of magnesium?

A $Mg^+(s) \longrightarrow Mg^{2+}(g) + e^-$

B $Mg^+(g) \longrightarrow Mg^{2+}(g) + e^-$

C $Mg(s) \longrightarrow Mg^{2+}(s) + 2e^-$

D $Mg(s) \longrightarrow Mg^{2+}(g) + 2e^-$

E $Mg(g) \longrightarrow Mg^{2+}(g) + 2e^-$

22. The successive ionization energies in kJ/mol of an element X are 740, 1500, 7700, 10 500, 13 600, 18 000, 21 700. Which ion is the most likely to be formed when X reacts with chlorine?

A X^{2-}

B X^-

C X^+

D X^{2+}

E X^{3+}

23. The electronic structures of argon, calcium, iron, selenium (Group 6) and one other element are given below (not in order). Which one represents the unnamed element?

A $1s^2 2s^2 2p^6 3s^2 3p^6 4s^2$

B $1s^2 2s^2 2p^6 3s^2 3p^6 3d^6 4s^2$

C $1s^2 2s^2 2p^6 3s^2 3p^6$

D $1s^2 2s^2 2p^6 3s^2 3p^6 3d^{10} 4s^2$

E $1s^2 2s^2 2p^6 3s^2 3p^6 3d^{10} 4s^2 4p^4$

24.

The mass spectrum of a metal is shown above. No other peaks were observed at other settings of mass number. The relative atomic mass of the metal is

A 63·2

B 63·4

C 63·6

D 63·8

E 64·0

25. The relative molecular mass of ^{12}CO is 27.994 9. The relative molecular mass of $^{12}CO_2$ is

    A 43.994 9

    B 43.989 8

    C 47.984 7

    D 39.994 9

    E some other number

26. When an element is bombarded with high-energy electrons, it emits X-rays the frequency of which is related to its

    A atomic number   D electron affinity

    B relative atomic   E ionization energy
       mass

    C mass number

27. Which pair of atomic numbers represents elements which are BOTH p-block elements?

    A 4, 8   D 10,20

    B 6,12   E 12,24

    C 8,16

28. The first ionization energies of five successive elements in the Periodic Table are 1400, 1310, 1680, 2080, 494 kJ mol^{-1}. These elements could be

    A the first five elements of a period

    B the last five elements of a period

    C all d-block elements

    D the last four elements of one period and the first element of the next period

    E the last element of one period and the first four elements of the next period

29. Which one of the following elements has a first ionization energy higher than that of carbon?

    A Boron   D Magnesium

    B Sodium   E Silicon

    C Neon

30. The table gives the first four ionization energies, in kJ mol^{-1}, of five elements (the letters are not the symbols for the elements). Which element occurs in Group 3 of the Periodic Table?

	First ionization energy	Second ionization energy	Third ionization energy	Fourth ionization energy
A	502	4569	6919	9550
B	526	7305	11822	——
C	584	1823	2751	11584
D	596	1152	4918	6480
E	793	1583	3238	4362

31. Which of the following ions would undergo the greatest deflection in a mass spectrometer?

    A $^{35}Cl^+$ D $^{35}Cl^{37}Cl^+$

    B $^{37}Cl^+$ E $^{35}Cl^{37}Cl^{2+}$

    C $^{35}Cl^{2+}$

Questions 32–36 concern the table below which shows the successive molar ionization energies (kJ mol^{-1}) of five elements Q to U.

The letters are not the symbols of the actual elements concerned.

Element	Ionization energy			
	1st	2nd	3rd	4th
Q	2 080	4 000	6 100	9 400
R	500	4 600	6 900	9 500
S	740	1 500	7 700	10 500
T	580	1 800	2 700	11 600
U	420	3 100	4 400	5 900

32. Which two elements are most likely to be in the same group of the Periodic Table?

    A Q and R

    B S and T

    C T and U

    D R and T

    E R and U

33. Which is most likely to be the correct equation for discharge at the cathode during the electrolysis of the fused halide?

    A $Q^{2+} + 2e^- \rightarrow Q$

    B $R^{2+} + 2e^- \rightarrow R$

    C $S^{3+} + 3e^- \rightarrow S$

    D $T^{3+} + 3e^- \rightarrow T$

    E $U^{2+} + 2e^- \rightarrow U$

34. Which is most likely to be the correct formula for the chloride of the element concerned?

    A QCl_2

    B RCl

    C SCl_3

    D TCl

    E UCl_4

35. Element S is most likely to be

    A an *s*-block element

    B a noble (or inert) gas

    C a *p*-block element

    D a metalloid

    E a *d*-block element

36. Which element is chemically and physically most like element Q?

    A Boron ($1s^2 2s^2 2p^1$)

    B Beryllium ($1s^2 2s^2$)

    C Lithium ($1s^2 2s^1$)

    D Hydrogen ($1s^1$)

    E Helium ($1s^2$)

Test 3   Answers

1. The outer electronic configuration of a noble gas is ns^2np^6 with the s and p levels full. C is such an element. Which noble gas is it?

2. The element described is likely to be a transition element (d block). Look for an electronic configuration with two electrons in the outer s level (occasionally one) while the inner d level is incomplete. The element is **A**. Which transition element is **A**? Have you met a reaction which the oxide of **A** catalyses?

3. On going down a Group in the Periodic Table, the first ionization energy decreases. This is because increasing distance from the nucleus, which means decreasing attraction for the outer electron, is more important than increasing nuclear charge, which means increasing attraction for the outer electron. The first element in a Group has the highest first ionization energy of the elements in that Group.

   To answer question 3 it is necessary to find the electronic configuration which corresponds to the first element in a Group. This is an interesting problem as decision has to be made between **C** (neon) and **E** (lithium). Some Periodic Tables show both these elements as the first in their Groups while other Tables show hydrogen as the first member of Group 1 and/or helium as the first member of Group 0.

   It is reasonable to regard helium as the first member of Group 0 as its properties, or lack of properties, place it among the noble gases. Hydrogen, on the other hand, is in the unique position of being both the first (Group 1) and penultimate (Group 7) member of its Period. Hydrogen, like the alkali metals, has only one outer electron but, like the halogens, needs only one more electron to have an electronic configuration resembling that of a noble gas.

   Some Periodic Tables put hydrogen at the top of Group 7 — if you try making a list of ways in which hydrogen does resemble the other halogens you are likely to find that hydrogen is better placed in Group 7 than in Group 1. The existence of the solvated hydrogen ion (such as H_3O^+) is the only reason at the moment for considering hydrogen as a member of Group 1. There is considerable speculation about the possible existence of metallic hydrogen. If this is ever found to exist, question 3 may require amendment but at the moment the best answer is **E**.

4. **1** is correct.

   2 is more difficult as it is almost correct but is made untrue by the description of the lines as converging at the *low* frequency end. The lines do converge but do so towards the high frequency end (*see* question 16).

   3 is correct, the convergence limit of the series of highest frequency, the Lyman series, providing one way of measuring the ionization energy of hydrogen.

   4 is incorrect as the spectrum described is concerned with atoms, not molecules. The correct key is **B**. The best wrong answer is **A**.

5. **1** is incorrect. It sounds plausible but is a reference to Moseley's experiment to determine the atomic number of an element.

   2 is correct. The conductivity of the vapour shows a sudden change as the energy of the bombarding electrons becomes just sufficient to persuade each atom of the vapour to eject its own outer electron.

   3 is incorrect. The method is used to find bond lengths and bond angles in molecules of vapour.

   4 is correct. This is the convergence limit method mentioned in question 4.

   2 and **4** are correct so the key is **C**.

6. This question is concerned with the equation

   Energy = Planck's constant × frequency

   1 and **3** are correct as the product of the two quantities mentioned will give the ionization energy in the required units.

   2 would be correct if Planck's constant is given per molecule rather than per mole. Conversion to the units required by the question would then involve multiplication by the Avogadro constant.

4 is incorrect. It is the position of one particular point (the convergence limit) in the spectrum which is used to find the first ionization energy. The only way in which 'number' comes into the determination is that investigation must be made on the convergence limit of 'series number 1' which is the first (Lyman) series. Only **1** and **3** are correct so key is **B**.

7. **1** is correct. The complete plot gives the best visual evidence available that electrons are arranged in energy 'shells'.

   2 and **3** are both correct. The energy jump between electrons 2 and 3 shows that there are two relatively easily removed electrons so the element is in Group 2. The appearance of the fifth electron in the graph shows that the Group 2 element cannot be beryllium which only has a total of four electrons. So the element must be magnesium, atomic number 12, or an element in Group 2 with atomic number more than 12.

   4 is incorrect as the rise in ionization energy shown in the graph is connected with *decreasing* distance of the electron from the nucleus. The rise is also connected with a decrease in repulsion from the remaining electrons as these become fewer in number. The correct key is **A**.

   Note that the scale on the vertical axis is logarithmic. The numerical values for the successive ionization energies of a Group 1 element will be found in question 22. The values in question 22 show more clearly the large proportional increase in ionization energy after removing the second electron.

8. This question can be answered most rapidly by considering the total number of electrons rather than the electronic configuration. Argon has 18 electrons so the configuration shown corresponds to 24 electrons. Which of the ions shown have this number of electrons? You need to know the order in which elements in the Periodic Table follow argon. This is K, Ca, Sc, Ti, V, Cr, Mn(25), Fe(26), Co(27), Ni(28). The numbers shown are the atomic numbers which a are also the numbers of electrons in neutral atoms. The numbers of electrons in the problem ions are Mn(23), Fe(23), Co(24) and Ni(26). Only Co(III) has the required number of electrons so the key is **E**.

9. This is another question about particles having the same total number of electrons. Such particles are called 'isoelectronic'. The electronic configuration given is that of argon, 2.8.8. Atoms of elements just before argon in the Periodic Table can move to this electronic configuration by electron gain. Sulphur, 2.8.6, and chlorine, 2.8.7, do this when they form the ions S^{2-} and Cl^{-}. Atoms of elements just after argon can move to this electronic configuration by electron loss. Calcium, 2.8.8.2, and scandium, 2.8.9.2, do this when they form the ions Ca^{2+} and Sc^{3+}. The ion Br^{-} has the configuration of another noble gas, krypton, which is achieved by electron gain.

   1, **2** and **3** are correct so the key is **A**.

10. The second ionization energy involves the change marked **IV** in the following sequence:

$$M(s) \xrightarrow{\text{I}} M(g) \xrightarrow[\;+\,e^-]{\text{III}} M^{+}(g) \xrightarrow[\;+\,e^-]{\text{IV}} M^{2+}(g)$$

$$M(g) \xrightarrow{\text{V}} M^{2+}(g) + 2e^-$$

$$M(s) \xrightarrow{\text{II}} M^{2+}(g) + 2e^-$$

   None of the possibilities from which you can choose is just **IV**. Of those given, both **2**, (**V** − **III**), and **4**, (**II** − **I** − **III**), produce the second ionization energy by difference so the correct key is **C**. This is a difficult question. An easier version of this problem is found in question 21.

11. β-Decay involves loss of an electron from the nucleus. The electron is of negligible relative atomic mass so the mass number remains at 212 and **1** is correct. Loss of one negative charge from the nucleus is equivalent to the nucleus becoming more positive by one unit. The atomic number goes up by unity to 83 so **2** is incorrect. At the end of the first ten hours, half the original lead will remain and at the end of the next ten hours half of this will have gone. The fraction remaining after 20 hours will be $\frac{1}{2} \times \frac{1}{2}$ which is $\frac{1}{4}$ so **3** is correct. It is not possible to tell from the information given whether the

nucleus formed is stable or unstable so **4** is incorrect. **1** and **3** are correct and the key is **B**.

12. **1** is correct. On going down any Group in the Periodic Table, first ionization energy decreases as the first electron to be removed becomes further away from the control of the nucleus and more influenced by repulsion of the steadily increasing number of other electrons.
    2 is correct. The configuration

    Noble gas s^1

    distinguishes an alkali metal from all other elements.
    The chemical reactivity of the alkali metals does increase with increasing atomic number so **3** is correct. Probably the most important reason for this reaction order is connected with the melting points of the alkali metals. Look up the melting points in a data book and try to explain the observed reactivities. Atoms of alkali metals lose their unpaired s electron and move to the electronic configuration of a noble gas. **4** is correct. All four statements are correct so the key is **E**.

13. The best description of what Geiger and Marsden found is that given in **E**. In this experiment, most of the α-particles went through almost undeflected. These α-particles were passing through the portion of the atom occupied by the electrons. The occasional α-particle passed close to a nucleus and was deflected, sometimes through a large angle. The small proportion of these deflections indicated that the volume of the nucleus is very small compared with the total volume of the atom. It is difficult to visualize how small, relatively, the nucleus is. To try to obtain some idea, imagine the atoms in the metal foil scaled up so that each nucleus is the size of a football. On this model the nuclei are about 20 miles apart. The rest of the volume of the atom is filled, very thinly, by the electrons.

14. The convention is to write

    $^{\text{mass number}}_{\text{atomic number}}$ Symbol

    The mass number of sodium is 23 which is the sum of the numbers of protons and neutrons.
    The atomic number of sodium is 11 which is the number of protons. So the number of neutrons is $23 - 11$ which is 12 and the key is **C**.

15. Neutron bombardment is an important way of converting one element into another. When you meet a question of this type you need to remember two things:
    (1) The sum of the mass numbers on both sides of the equation is equal. (These are the top numbers.)
    (2) The sum of the atomic numbers (nuclear charge numbers) on both sides of the equation is equal. (These are the lower numbers.)

    If the particle X is written as $^m_c X$,

    $$27 + 1 = m + 4 \quad \text{so } m = 24$$
    $$13 + 0 = c + 2 \quad \text{so } c = 11$$

    B is the correct key.

16. The emission spectrum of atomic hydrogen is composed of several series of lines, the lines in each series converging towards the direction of higher frequency. This is happening in **E**. **C** is the best incorrect answer. Decision between **C** and **E** had to be made when answering question 4.

17. The second series, called the Balmer series, is the one observed in the visible region of the spectrum. It is produced by electron energy drops from higher levels to level 2. The drop from level 3 to level 2 produces the first (red) line, 4 to 2 produces the second line, 5 to 2 the third and 6 to 2 the fourth which is in the violet region. The correct key is **C**.

18. If you were short of time and had to make an intelligent chemical guess to answer the question, you would look for an electronic structure in which outer s electrons had been lost and as many electrons as possible were unpaired. You would have to guess between **B** and **D** and would have a much greater chance of being correct than if you had guessed between all five possibilities. To answer with certainty, decide how many electrons uncharged iron atoms possess in excess of argon. The order of the elements beyond argon shows that this number is 8 (*see* question 8).
    The Fe(III) ion will therefore possess five electrons in excess of argon. These electrons will be as unpaired as possible so the correct key is **D**.

19. You need to be familiar with the Periodic variation in the first ionization energy to answer this question. The general pattern is one of increase across a period followed by a large drop from the noble gas to the next alkali metal, after which the pattern is repeated. *M, Q, R* and *T* are at the peaks in the graph before the drop to the alkali metal so they are noble gases and the key is **E**.

What is the reason for the increase in first ionization energy on going across a period towards the noble gas and why is there a sudden drop in value after the noble gas?

20. The elements are hydrogen to boron. If you are familiar with the pattern discussed in question 19, you will know that noble gases have high first ionization energies and that helium has the highest of all elements. The correct key is **B**.

21. This is a far easier problem than question 10. You only need to know that the second ionization energy involves loss of the second electron from a gaseous ion of charge +1. The correct equation is **B**.

22. Look for the position where there is a large *proportional* increase in the ionization energy. For X this occurs between 1500 and 7700, a five fold increase. Loss of two electrons is relatively easy while loss of the third is far more difficult. The loss of the third electron must involve removing an electron from an inner shell. X has two outer electrons so it is in Group 2 and will form 2+ ions. The key is **D**.

23. Look for the electronic structures of the named elements. Argon, a noble gas, will have the outer configuration $ns^2 np^6$. This is **C**. Calcium, a Group 2 element coming before the filling of the *d* levels, will have an outer configuration $np^6 (n+1)s^2$. This is **A**. Iron, a transition element somewhere in the middle of the sequence, will have an incomplete *d* level while the outer *s* level usually contains two electrons. This is **B**. Selenium, in Group 6, will have the outer configuration $ns^2 np^4$. This is **E**. The remaining element is **D** which is the correct key. Which element is this?

24. The relative atomic mass will be the weighted mean of the two mass numbers involved, 63 and 65. The abundance ratio of these isotopes is 7 to 3

$$\text{RAM} = 63 \times (7/10) + 65 \times (3/10) = 63 \cdot 6$$

The key is **C**.

25. The contribution made by carbon to the relative molecular mass of ^{12}CO is $12 \cdot 0000$. This means that the relative atomic mass of the oxygen isotope involved in the combination is $15 \cdot 9949$.

$$\text{RMM of } ^{12}CO_2 = 12 \cdot 0000 + 2 \times 15 \cdot 9949$$
$$= 43 \cdot 9898$$

The key is **B**.

26. This is a description of Moseley's experiment in which a relationship was found between the x-ray frequency and the atomic number of the element concerned, key **A**. Can you find in a text-book the equation connecting x-ray frequency and the atomic number? What functions of these two quantities are plotted in order to obtain a straight-line graph?

27. The pairs of elements with the atomic numbers given are

s-block	*p*-block	*d*-block
A (4) beryllium	(8) oxygen	
B (12)magnesium	(6) carbon	
C	(8) oxygen (16)sulphur	
D (20)calcium	(10)neon	
E(12)magnesium		(24)chromium

The *p*-block pair are oxygen and sulphur, Key **C**.

28. You need to be familiar, as in answering question 19, with the way in which the first ionization energy varies with atomic number in a periodic way. The large drop in value from 2080 to 494 kJ mol^{-1} locates the break between a noble gas and the alkali metal of the next period. The Key is **D**.

29. Once again, you need to know how the first ionization energy changes on moving across a period and on going from one period to the next. The change across a period is one of general increase. (See the graph in question 19). Boron will have a lower value than carbon, while neon's value will be greater than that of carbon. This shows that the key is **C**. The large drop in ionization energy after neon means that the values for sodium, magnesium and silicon are all lower than for carbon. Are there any elements in the period starting with sodium that have a higher first ionization energy than carbon?

30. A Group 3 element will require steadily increasing energies to remove the first three electrons from its gaseous atoms. The fourth electron will have to be removed from an inner shell and the fourth ionization energy will show a large proportional increase in value. This is the case in **C** where the energy required to remove the fourth electron is over four times the energy needed to remove the third electron.

31. The deflection in a mass spectrometer will be greatest for ions of small mass and high charge. Both these factors are operating to make key **C** the correct choice.

32. It is best to decide in which Group each element is placed. This will help in answering all the questions in the set. Look for the point where there is a large proportional increase in ionization energy. This tells you when electrons begin to be removed from inner shells.
    The Group numbers are:

Q   Group 4 or beyond. There is no large proportional increase within the values given

R   Group 1 as there is a nine-fold increase after the first electron has been removed

S   Group 2

T   Group 3

U   Group 1

R and U are both in Group 1 so the key is **E**.

33. Decide what charge will be carried by the positive ions of each element. Q may be a non-metal and form no positive ions. But if it is in Group 4, and there is no firm evidence for this, it could form Q^{2+} ions in the same way that lead and tin do. **A** could be correct but is there an equation that is more certain to be correct?
    R and U will form 1+ ions so **B** and **E** are incorrect. S will form a 2+ ion so **C** is incorrect.
    T is in Group 3 and will form 3+ ions unless it is the non-metal boron. These ions could be discharged at the cathode during electrolysis of the molten halide. There is positive evidence that **D** could be right so this would be chosen as the equation 'most likely to be correct'.

34. The formula of the chloride of Q will not be certain. There is no doubt about the chlorides of the Group 1 elements, RCl and UCl. The Group 2 element will form a chloride SCl$_2$. The Group 3 element is most likely to form a chloride TCl$_3$ but it could, if the element was sufficiently low in the Group, also form a chloride TCl. This is an example of the general pattern that the possible valencies (or oxidation numbers if you have met the term) are the Group number or multiples of 2 less than the Group number.
    Coming back to the decision about what is 'most likely to be the correct formula', there is no doubt that the choice has to be **B** as there is no doubt about the formula RCl.

35. S is in Group 2 so it is an *s*-block element, key **A**.

36. A teasing problem to finish the test! There is no sign in the first four ionization energies of where the jump in value is going to come. It could come after the fourth electron has been removed or it could come later. This means that Q is in one of the Groups 4, 5, 6, 7 or 0.

Can you find an element in the five possibilities with 4, 5, 6, 7 or 8 electrons in its outer shell? No! Consider the wording of the question again. You are trying to find an element which is 'chemically and physically most like' an element in Group 4, 5, 6, 7 or 0. The noble gas helium is the best choice even though it does not have 8 electrons in its outer shell. The key is **E**.

The next-best choice is **D** as hydrogen does show certain resemblances to the halogens and Q could be a halogen. Helium, however, resembles the other noble gases more closely than hydrogen resembles the halogens. You might like to read again the final two paragraphs of the answer to question 3.

Test 4

Acid–base, redox, oxidation numbers, titrations

Questions 1–4 refer to the titration of a chloride solution with silver nitrate solution.

The student measured the chloride solution into a conical flask with a pipette, added chromate indicator and then ran in the silver nitrate solution from a burette. In each question below, decide whether the situation described would

A make his estimate of the chloride concentration too high

B make his estimate of the chloride concentration too low

C involve an error which reduced the accuracy of his determination in a random way

D involve errors which cancel each other out

E involve no error of any sort

1. The student did not always read the burette with his eye at the level of the meniscus.

2. The jet of the burette was not initially filled with the silver nitrate solution.

3. There was distilled water left in the conical flask after each wash and rinse between titrations.

4. At the beginning of the experiment the student had poured the silver nitrate solution into a beaker which he had carefully washed with distilled water only.

Directions summarized for questions 5 to 13				
A	**B**	**C**	**D**	**E**
1,2,3 only correct	**1,3** only correct	**2,4** only correct	**4** only correct	Some other response or combination of responses is correct

5. In which of the following ions does the metal have an oxidation number of +3?

   1 VO^{2+}

   2 AlO_2^-

   3 $[Fe(CN)_6]^{4-}$

   4 $[CrCl_2(H_2O)_4]^+$

6. In which of the following is there an element with the same oxidation number as that of chromium in $K_2Cr_2O_7$?

   1 Cl_2O_7

   2 $Fe(CN)_6^{3-}$

   3 VO_2^+

   4 K_2MnO_4

7. In which conversions is chlorine oxidized?

   1 $Cl_2 \rightarrow OCl^-$

   2 $Cl_2 + I_2 \rightarrow 2ICl$

   3 $OCl^- \rightarrow ClO_3^-$

   4 $OCl^- \rightarrow Cl^-$

8. In which of the following reactions is the named element oxidized?

   1 Thallium in
      $$2Tl^+(aq) + Zn(s) \rightarrow$$
      $$2Tl(s) + Zn^{2+}(aq)$$

   2 Silver in
      $$Ag_2S(s) + 4CN^-(aq) \rightarrow$$
      $$2[Ag(CN)_2]^-(aq) + S^{2-}(aq)$$

   3 Copper in
      $$2Cu^{2+}(aq) + 4I^-(aq) \rightarrow$$
      $$2CuI(s) + I_2(aq)$$

   4 Gold in
      $$4Au(s) + 8CN^-(aq) + 2H_2O(l) + O_2(g) \rightarrow$$
      $$4[Au(CN)_2]^-(aq) + 4OH^-(aq)$$

9. Which equation(s) might be used to illustrate the action of nitric acid as an oxidant?

   1 $P_4O_{10} + 4HNO_3 \rightarrow 4HPO_3 + 2N_2O_5$

   2 $6Fe^{2+} + 8HNO_3 \rightarrow$
      $\qquad 6Fe^{3+} + 2NO + 4H_2O + 6NO_3^-$

   3 $CO_3^{2-} + 2HNO_3 \rightarrow CO_2 + H_2O + 2NO_3^-$

   4 $Cu + 4HNO_3 \rightarrow$
      $\qquad Cu^{2+} + 2NO_3^- + 2H_2O + 2NO_2$

10. Chlorine can exist with oxidation numbers ranging from -1 to $+7$. In which of the following ions could chlorine NOT undergo disproportionation?

   1 ClO^-

   2 ClO_4^-

   3 ClO_3^-

   4 Cl^-

11. To which of the types below does the given reaction belong?

   $$Ba(OH)_2(aq) + H_2C_2O_4(aq) \rightarrow$$
   $$2H_2O(l) + BaC_2O_4(s)$$

   1 Oxidation–reduction

   2 Precipitation

   3 Disproportionation

   4 Neutralization

12. True statements about a solution of dry hydrogen chloride gas in dry toluene include that the

   1 solution does NOT fume in moist air

   2 hydrogen chloride reacts with the toluene

   3 solution turns blue litmus red

   4 solution is a non-conductor of electricity

13. Which of the following are acid–base conjugate pairs?

   1 HCO_3^- and CO_3^{2-}

   2 NH_4^+ and NH_2^-

   3 HCl and Cl^-

   4 H_3O^+ and OH^-

Directions for questions 14 to 21. Each of the questions or incomplete statements in this section is followed by five suggested answers. Select the best answer in each case.

14. In which of the following reactions has the NAMED substance NOT acted as an acid?

   A $HSO_4^-(aq) + NH_3(g) \rightarrow NH_4^+(aq) + SO_4^{2-}(aq)$
   Hydrogen sulphate ion

   B $NH_4^+(aq) + NaNH_2(s) \rightarrow Na^+(aq) + 2NH_3(g)$
   Ammonium ion

   C $Mg_3N_2(s) + 6H_2O(l) \rightarrow$
   $\qquad 3Mg(OH)_2(s) + 2NH_3(g)$
   Water

   D $CaCO_3(s) + 2CH_3CO_2H(aq) \rightarrow$
   $Ca^{2+}(aq) + 2CH_3CO_2^-(aq) + CO_2(g) + H_2O(l)$
   Acetic (ethanoic) acid

   E $NaH(s) + H_2O(l) \rightarrow$
   $\qquad H_2(g) + Na^+(aq) + OH^-(aq)$
   Sodium hydride

15. The number of cm^3 of 0·25M HBr required to neutralize 100 cm^3 of 0·125M $Ba(OH)_2$ solution is.

   A   25

   B   50

   C   75

   D 100

   E 125

16. 17·1 g of aluminium sulphate, $Al_2(SO_4)_3$, was dissolved in water and made up to one dm^3 of solution.
   What was the concentration (in mol dm^{-3}) of the solution with respect to sulphate ions?
   $[Al_2(SO_4)_3 = 342]$

   A 0·005

   B 0·0167

   C 0·05

   D 0·15

   E 0·25

17. Select the reaction in which the oxidation number of nitrogen changes from +5 to +1.

   A $3Cu(s) + 12H^+(aq) + 6NO_3^-(aq) \rightarrow$
   $\qquad 3N_2O_4(g) + 6H_2O(l) + 3Cu^{2+}(aq)$

   B $3Cu(s) + 8H^+(aq) + 2NO_3^-(aq) \rightarrow$
   $\qquad 2NO(g) + 4H_2O(l) + 3Cu^{2+}(aq)$

   C $3Cu(s) + 7·5H^+(aq) + 1·5NO_3^-(aq) \rightarrow$
   $\qquad 0·75N_2O(g) + 3·75H_2O(l) + 3Cu^{2+}(aq)$

   D $3Cu(s) + 7·2H^+(aq) + 1·2NO_3^-(aq) \rightarrow$
   $\qquad 0·6N_2(g) + 3·6H_2O(l) + 3Cu^{2+}(aq)$

   E $3Cu(s) + 7·5H^+(aq) + 0·75NO_3^-(aq) \rightarrow$
   $\qquad 0·75NH_4^+(aq) + 2·25H_2O(l) + 3Cu^{2+}(aq)$

18. A commercial production of iodine involves the reduction of a solution of iodate(V) ions (IO_3^-) with the theoretical quantity of hydrogen sulphite ions (HSO_3^-). The hydrogen sulphite is oxidised to sulphate ions (SO_4^{2-}) while the iodate(V) is reduced to iodine (I_2). How many moles of hydrogen sulphite ions are needed to reduce one mole of iodate(V) ions?

   A 0·4

   B 1

   C 2

   D 2·5

   E 5

19. Given the equations

$$S_2O_8^{2-} + 2e^- \rightarrow 2SO_4^{2-}$$

$$Mn^{2+} + 4H_2O \rightarrow MnO_4^- + 8H^+ + 5e^-$$

how many moles of $S_2O_8^{2-}$ ions are required to oxidize 1 mole of Mn^{2+} ions?

A 0·4

B 0·5

C 1·0

D 2·0

E 2·5

20. 20 cm³ of a 0·1 M solution of metal ions reacted with 20 cm³ of 0·1 M sulphur dioxide solution. The sulphur dioxide reacted according to the equation

$$SO_2(aq) + 2H_2O(l) \rightarrow$$
$$SO_4^{2-}(aq) + 4H^+(aq) + 2e^-$$

If the original oxidation number of the metal was +3, the new oxidation number of the metal would be

A   0

B +1

C +2

D +4

E +5

21. In which compound does vanadium have an oxidation number of +4?

A NH_4VO_2

B $K_4[V(CN)_6]$

C VSO_4

D $VOSO_4$

E VCl_3

22. In which one of the following reactions does the FIRST mentioned species act as a base?

A $H_2O(l) + NH_3(g) \rightarrow NH_4^+(aq) + OH^-(aq)$

B $H_3O^+(aq) + OH^-(aq) \rightarrow 2H_2O(l)$

C $HCl(g) + H_2O(l) \rightarrow H_3O^+(aq) + Cl^-(aq)$

D $CH_3COCH_3(aq) + HCl(g) \rightarrow$

$$CH_3 \overset{+}{C}CH_3(aq) + Cl^-(aq)$$
$$|$$
$$OH$$

E $C_2H_5OH + CH_3CO_2H \rightarrow$
$$CH_3CO_2C_2H_5 + H_2O$$

23. How many moles of zinc manganate (VII), $Zn(MnO_4)_2$, are needed to react completely with 30 cm³ of acidified 0.1 M iron (II) sulphate solution?

$$MnO_4^- + 8H^+ + 5e^- \rightarrow Mn^{2+} + 4H_2O$$
$$Fe^{2+} \rightarrow Fe^{3+} + e^-$$

A 3×10^{-2} D 6×10^{-4}

B 7.5×10^{-3} E 3×10^{-4}

C 1.2×10^{-3}

24. One mole of hydrazine, N_2H_4, reacts with one mole of selenic (IV) acid, H_2SeO_3; the latter is reduced to selenium(0).

Into which of the following is hydrazine changed?

A N_2 D $N_2H_5^+$

B $2NH_3$ E $NH_3 + \frac{1}{2}N_2$

C $2NH_2OH$

25. In an experimental investigation of the reduction of chlorate (VII) ion (ClO_4^-) in aqueous solution, it was found that 25.0 cm³ of 0.0500 M aqueous potassium chlorate (VII) required 50.0 cm³ of 0.200 M aqueous titanium (III) chloride to reach the end-point. (The titanium(III) ion is oxidized to

titanium (IV) ion in this reaction.)
Which one of the following formulae could correctly represent the reduction product of the chlorate (VII) ion?

A Cl_2 C OCl^- E ClO_3^-

B Cl^- D ClO_2^-

26. 0.9 g of a metal oxide MO were dissolved in excess dilute sulphuric acid. 25 cm^3 of 0.1 M potassium manganated(VII) were required to oxidize this solution.

$$M^{2+} \rightarrow Mn^{2+} + e^-$$
$$MnO_4^- + 8H^+ + 5e^- \rightarrow Mn^{2+} + 4H_2O$$

If O = 16, the relative atomic mass of M is

A 20   D 56

B 28   E 72

C 40

Questions 27–30 concern an experiment to determine the quantity of potassium iodide in a sample of table salt.

5 g of the salt was dissolved in water and the solution acidified with sulphuric acid. To this was added an excess of bromine water, and the resulting solution was boiled until all the iodide present had been converted into iodate, IO_3^-. After boiling off all excess bromine, the solution was cooled and then treated with a small quantity of potassium iodide. The iodine liberated in the reaction between iodide and the iodate previously formed was titrated with 0.005M sodium thiosulphate solution, in the presence of starch indicator. 12.0 cm^3 of the thiosulphate solution were required to discharge the colour in the flask.

27. What is the numerical *change* in the oxidation number of iodine during the oxidation of the iodide to iodate?

A 1   D 6

B 4   E 7

C 5

28. Excess bromine was removed by boiling. If a larger excess had been present one of the following substances could have been used to remove it. Which?

A Sodium bromate

B Sodium thiosulphate

C Silver nitrate

D Dilute acetic acid

E Phenol

29. How many moles of iodine molecules (I_2) are produced by one mole of potassium iodate in this determination?

A 0.60

B 1.67

C 3.00

D 5.00

E 6.00

30. If 2 moles of sodium thiosulphate are equivalent to each mole of iodine molecules (I_2) liberated, how many moles of potassium iodide* were originally present in the table salt sample?

(*Remember this was converted to potassium iodate.)

A 1.0×10^{-6}

B 3.6×10^{-6}

C 1.0×10^{-5}

D 1.2×10^{-5}

E 3.6×10^{-5}

Test 4 Answers

1. Failure to read the burette with the eye at menicus level will give a volume reading which is either too large or too small. The magnitude and direction of the error will depend on how far the angle of the eye is from the correct position.

   The student's value for the volume of silver nitrate solution will be given by the difference between two readings each of which can, independently, be either too large or too small. This random error is described in key **C**.

2. If the burette jet is not filled before titration is begun, some silver nitrate solution will be used to fill the jet before any can run into the conical flask. The student's value for the volume of silver nitrate apparently required to react with the solution of chloride will be too large. This will make the chloride concentration appear to be higher than it really is. The correct key is **A**.

   If the student were to repeat the titration several times, topping up the burette when necessary, the correct key would not be **A**. The first reading would be rejected as a rough value and only later readings would be used. What would the correct key be under these conditions?

3. The number of moles of chloride taken in the experiment is determined only by the volume of chloride solution run from the pipette. The presence of water in the flask has no influence on the number of moles of chloride or the volume of silver nitrate required. The correct key is **E**.

4. Pouring the silver nitrate into a damp beaker will slightly dilute the solution so that a greater volume than the correct value will be required in the titration. The situation is similar to that in question 2 and the key is **A**.

5. To answer this question you need to know the other oxidation numbers. O is nearly always -2 (*see* question 19), CN carries a charge of -1 (so it is not necessary to know the individual oxidation numbers of C and N here), H_2O is a neutral molecule within which the oxidation numbers cancel and Cl is -1 in chlorides.

   You also need to know that the sum of the oxidation numbers is zero for a neutral compound or equal to the value of the charge on an ionized particle.

   All the oxidation numbers have been inserted above the formulae, including the value the metal must have to give the correct charge on the ion.

$$\overset{+4\ -2}{V\ O}\ \ \ \overset{2+}{} \qquad \overset{+3\ -4}{Al O_2}\overset{1-}{} \qquad \left[\overset{+2\ \ -6}{Fe(CN)_6}\right]^{4-}$$

$$\overset{+3\ -2\ \ \ \ 0}{[CrCl_2(H_2O)_4]}\overset{1+}{}$$

The oxidation number of the metal is $+3$ in **2** and **4** so the correct key is **C**.

6. K is $+1$, O is -2 and the whole compound is neutral. Cr must have an oxidation number of $+6$.

$$\overset{+6}{}$$
$$\overset{+2\ \ +6\ \ -14}{K_2\ Cr_2\ O_7}$$

Using $O = -2$, $CN = -1$ and $K = +1$, the oxidation numbers of the other elements can be calculated:

$$\overset{+7}{}$$
$$\overset{+7\ \ -14}{Cl_2\ O_7} \qquad \left[\overset{+3\ \ -6}{Fe(CN)_6}\right]^{3-} \left[\overset{+5\ -4}{V\ O_2}\right]^{1+}$$

$$\overset{+2\ \ +6\ \ -8}{K_2\ Mn\ O_4}$$

Only **4**, key **D**, contains an element with oxidation number $+6$.

   There is a problem over CN^-. Does one of the elements here have an oxidation number of $+6$? The range of oxidation numbers for an element run from $+$ its Group number in the Periodic Table to 8 less than this. C has values from $+4$ to -4 while N has values from $+5$ to -3. Neither of these ranges include $+6$.

7. The changes in oxidation number for chlorine are

$$1, 0 \to +1 \quad 2, 0 \to -1 \quad 3, +1 \to 5 \quad 4, +1 \to -1$$

Oxidation is an increase in oxidation number. This is happening to chlorine in **1** and **3**, key **B**.

8. The changes in oxidation number are

1 $Tl(+1) \to Tl(0)$ reduction

2 $Ag(+1) \to Ag(+1)$ not redox

3 $Cu(+2) \to Cu(+1)$ reduction

4 $Au(0) \to Au(+1)$ oxidation

Only **4** is oxidation. This is key **D**.

9. An oxidant is an oxidizing agent. It is necessary to decide in which cases the substance reacting with nitric acid is being oxidized. The oxidation number changes in the substance reacting with nitric acid are

1. $\overset{4 \times +5}{P_4} \quad \overset{10 \times -2}{O_{10}} \quad \to \overset{+1}{H} \overset{+5}{P} \overset{3 \times -2}{O_3}$

2. $Fe(+2) \qquad \to Fe(+3)$

3. $\left[\overset{+4}{C} \overset{3 \times -2}{O_3} \right]^{2-} \quad \to \overset{+4}{C} \overset{2 \times -2}{O_2}$

4   $Cu(0) \qquad \to Cu(+2)$

Oxidation is taking place in **2** and **4**, key **C**.

10. Disproportionation of chlorine takes place when chlorine oxidizes and reduces itself. The element in a molecule or ion is both increased and reduced in oxidation number. For this to happen, chlorine must have an intermediate oxidation number so that this number can be increased and reduced.

   The oxidation numbers of chlorine (Group 7) run from +7 to -1. If chlorine has one of these oxidation numbers it cannot undergo disproportionation. The oxidation numbers of chlorine in the question are +1 in ClO^-, +7 in ClO_4^- +5 in ClO_3^- and -1 in Cl^-. Chlorine cannot undergo disproportionation in **2** and **4** so the correct key is **C**.

   The disproportionation reactions of the ions in **1** and **3** are given in the equations below. The oxidation numbers of chlorine have been put in above the formulae.

$$\overset{+1}{3ClO^-} \to \overset{-1}{2Cl^-} + \overset{+5}{ClO_3^-}$$

$$\overset{+5}{4ClO_3^-} \to \overset{-1}{Cl^-} + \overset{+7}{3ClO_4^-}$$

Care is needed in reading this question. Chlorine in the perchlorate ion [which may be called chlorate(VII) on your course] cannot disproportionate as chlorine is in its highest oxidation state. The whole ion can, however, disproportionate. The chlorine can oxidize the oxygen and the oxygen can reduce the chlorine. This happens when a perchlorate is heated in an ignition tube. Oxygen is evolved as it is oxidized from -2 to 0 in oxidation number.

$$\overset{+7}{Cl} \overset{4 \times -2}{O_4^-} \to \overset{-1}{Cl^-} \times \overset{4 \times 0}{2 O_2}$$

11. The state symbols in the equation show that precipitation of solid BaC_2O_4 is taking place from aqueous solution. The reaction is also a neutralization of the acid $H_2C_2O_4$ by the base $Ba(OH)_2$. The reaction does not involve oxidation or reduction. Only **2** and **4** are correct and the key is **C**.

12. A solution of HCl gas in toluene gives off HCl gas all the time. This gas fumes in moist air. **1** is incorrect.

   Dissolution of HCl in toluene must involve some form of interaction between molecules or it would not dissolve. But no new substances are formed so one would not say that the HCl 'reacts'. **2** is incorrect.

   The question emphasizes that everything is dry. No reaction with water can take place.

$$\underset{\text{acid 1}}{HCl(g)} + \underset{\text{base 2}}{H_2O(l)} \rightleftharpoons \underset{\text{acid 2}}{H_3O^+(aq)} + \underset{\text{base 1}}{Cl^-(aq)}$$

$H_3O^+(aq)$ ions are not present so the solution will have no acidic properties unless water is introduced from outside. Introduction of water would occur if the HCl solution was tested with an aqueous solution of an indicator which would then show its acidic colour. The implication in statement **3**, however, is that dry indicator paper is used. Under these conditions the indicator will not turn to its acidic colour. The statement is incorrect.

No ions are formed on dissolution, both HCl and toluene remaining as molecules. The solution will be a non-conductor. This is the only correct statement and the key is **D**.

13. HCl and Cl^- form an acid–base pair. So do H_3O^+ and H_2O. This can be seen below the chemical equation in the answer to question 12 where one pair is called 1 and the other 2. An acid loses its proton to a base and forms the base of its acid–base pair. These pairs differ by a proton. The substances in **1** and **3** differ by a proton, those in **2** and **4** do not. The key is **B**.

14. In **A**, **B**, **C** and **D** the named substance donates protons to a molecule or an ion contained in the second substance. In these four cases the named substance acts as an acid.

The four equations are written below with some changes made so that the real acid and base can be more clearly identified in each case. Decide which species is acid 1, base 1, acid 2 and base 2.

$$HSO_4^- + NH_3 \rightleftharpoons NH_4^+ + SO_4^{2-}$$
$$NH_4^+ + NH_2^- \rightleftharpoons NH_3 + NH_3$$
$$N^{3-} + 3H_2O \rightleftharpoons 3OH^- + NH_3$$
$$CO_3^{2-} + 2CH_3CO_2H \rightleftharpoons 2CH_3CO_2^- + H_2CO_3$$

The named substance in **E**, sodium hydride (or hydride ions), has accepted a proton from water and is therefore not an acid. Water is the acid, NaH is the base. The simplified form of the equation is

$$H^- + H_2O \rightleftharpoons H_2 + OH^-$$

This is an interesting reaction as it is the very special case of a reaction which is both acid–base and redox. Check the oxidation numbers in the equation to find what is oxidized and what is reduced.

15. The equation for the reaction is

$$2HBr + Ba(OH)_2 \rightarrow BaBr_2 + 2H_2O$$

The equation shows that two moles of HBr are required for every mole of $Ba(OH)_2$. Two of the many ways of achieving neutralization would be to use twice the volume of HBr of the same molarity as the $Ba(OH)_2$ or to use the same volume of HBr of twice the molarity of the $Ba(OH)_2$. It is the second alternative which is being used in this problem where the correct key is **D**.

16. The number of moles of $Al_2(SO_4)_3$ per litre (or dm^3) is $17 \cdot 1/342$. As each mole of compound gives three moles of sulphate ions, the number of moles of sulphate per litre is $3 \times 17 \cdot 1/342$. This is a little more than $50/350$ or $1/7$ so the correct key is **D**. [You might also see the correct solution by writing the ratio as $3 \times 17.1/2 \times 171$].

17. Nitrogen starts at oxidation number $+5$ in each case and finishes at $+4$, $+2$, $+1$, 0 and -3. The correct key is **C**.

18. The oxidation number equation for the reaction must be (unbalanced)

$$I(+5) + S(+4) \rightarrow I(0) + S(+6)$$

To balance the 'ups and downs' in oxidation number the balanced equation is

$$2I(+5) + 5S(+4) \rightarrow 2I(0) + 5S(+6)$$

It is not necessary to go further and write a balanced chemical equation. The oxidation number equation shows that two moles of IO_3^- react with five moles of HSO_3^-. This is equivalent to key **D**.

19. As the numbers of electrons involved in the ionic half-equations are given, it is possible to balance for electron loss and gain. If the first equation is scaled up five times and the second is scaled up twice, the number of electrons lost and gained will be ten:

$$5S_2O_8^{2-} + 10e^- \rightarrow 10SO_4^{2-}$$

$$2Mn^{2+} + 8H_2O \rightarrow$$
$$2MnO_4^- + 16H^+ + 10e^-$$

This shows that five moles of $S_2O_8^{2-}$ react with two moles of Mn^{2+}. The correct key is **E**.

The complete equation for the reaction is obtained by adding the two scaled-up half-equations. The $10e^-$ cancel and disappear from the equation. Try to write the complete equation.

It is very difficult to oxidize manganese (+2) to permanganate, MnO_4^-, in one step. The very powerful oxidizing agent, persulphate, $S_2O_8^{2-}$, is one of the few reagents that will do this. Silver (+1) ions must also be present to catalyse the reaction. It is believed that persulphate ions oxidize silver to Ag(+3) which is the real oxidizing agent for converting Mn(+2) to Mn(+7).

Another point of interest in this reaction is concerned with the oxidation numbers of sulphur and oxygen in the persulphate ion. One might expect the oxidation numbers to be

$$\begin{bmatrix} 2 \times +7 & 8 \times -2 \\ S_2 & O_8 \end{bmatrix}^{2-}$$

with sulphur (+7) and oxygen (+2). But sulphur is in Group 6 of the Periodic Table so its oxidation numbers can only run between +6 and −2. There must be something unexpected about the structure of the persulphate ion. This ion contains two types of oxygen, one type with oxidation number −2, the other with oxidation number −1. In the structure below the unmarked oxygens are all −2.

Oxygen has oxidation number −2 if attached only to atoms of elements less electronegative than itself, such as sulphur. When attached only to itself, in O_2, the oxidation number is 0. The unusual oxidation number of −1 arises because oxygen is attached by one bond to an element less electronegative than itself while the other bond is to itself. This also occurs in hydrogen peroxide and other peroxides. You will be familiar with the use of oxidation states to decide if oxidation and reduction is involved in a reaction and to help balance the equation for the reaction if redox is taking place. The detection of unexpected and unusual structures is another use of oxidation numbers.

20. The information given shows that the metal ions and SO_2 react in the ratio 1 mole to 1 mole. Sulphur increases in oxidation number from +4 to +6. To balance the 'ups and downs' in oxidation numbers, the metal must decrease in oxidation number from +3 to +1. The correct key is **B**.

The equation for the oxidation of SO_2 includes two electrons. How can this information be used to confirm that the metal ions are reduced from +3 to +1?

21. The oxidation numbers and charges on ions that must be known to solve the problem are: NH_4 is +1, O is −2, K is +1, CN is −1, SO_4 is −2 and Cl is −1. These values make the oxidation number of vanadium +3 in **A**, +2 in **B**, +2 in **C**, +4 in **D** and +3 in **E**. The correct key is **D**.

22. In the first four reactions in the question, a proton is transferred from one species to the other as the reaction moves from left to right. In **A**, the H_2O molecule is supplying a proton and behaving as an acid. This is also the case with the H_3O^+ ion in **B** and the HCl molecule in **C**.

In **D**, however, the first species is gaining a proton. The propanone molecule, CH_3COCH_3, is acting as a base and the key is **D**. Equation **E**, as written, does not show ethanol, C_2H_5OH, acting either as an acid or as a base.

23. 30 cm^3 of 0.1M iron (II) sulphate contain 3×10^{-3} mole of Fe^{2+}. The equations provided show that, for the transfer of one mole of electrons, 1 mole of Fe^{2+} reacts with only $\frac{1}{5}$ mole of MnO_4^-. So 3×10^{-3} mole of Fe^{2+} will react with 6×10^{-4} mole of MnO_4^-.

The formula of zinc manganate (VII) shows that one mole of the compound provides 2 mole of MnO_4^-. So 3×10^{-4} mole of the compound is needed to provide the 6×10^{-4} mole of MnO_4^- which is required to react with 3×10^{-3} mole of Fe^{2+}. This is key **E**.

24. The oxidation number of nitrogen in hydrazine, N_2H_4, is −2. The oxidation number equation for its reaction with H_2SeO_3 is

$$2 N(-2) + Se(+4) \rightarrow 2 N(x) + Se(0)$$

One selenium goes down 4 units in oxidation number so two nitrogens must go up 4 units between them. This would be the case if each nitrogen went up 2 units to nitrogen (0). This corresponds to nitrogen gas, N_2, and the key is **A**.

25. This a similar problem to that in question 24 but in this case it is first necessary to calculate the reacting ratio of moles from the titration figures. The numbers of moles involved are

Cl (+7) $0.05 \times 25/1000 = 1.25 \times 10^{-3}$ mole

Ti (+3) $0.20 \times 50/1000 = 10 \times 10^{-3}$ mole

This shows that 8 mole of Ti(+3) react with 1 mole of Cl(+7). The oxidation number equation is

$$Cl(+7) + 8\ Ti(+3) \longrightarrow Cl(x) + 8Ti(+4)$$

Eight titaniums go up 1 unit in oxidation number so one chlorine must go down 8 units. This indicates that the change is Cl(+7) to Cl(−1). Chlorine is in oxidation state −1 in the chloride ion, Cl^-, and the key is **B**.

26. The oxidation of M^{2+} used $0.1 \times 25/1000$ mole of MnO_4^-. This is 2.5×10^{-3} mole of MnO_4^-. The equations given show that, for the transfer of 5 moles of electrons, 1 mole of MnO_4^- will oxidize 5 mole of M^{2+}. So 2.5×10^{-3} mole of MnO_4^- will oxidize 12.5×10^{-3} mole of M^{2+}.

The 0.9 g of MO that was used must contain 12.5×10^{-3} mole of M. The mass of MO that would contain 1 mole of M is $0.9 \times 1/12.5 \times 10^{-3}$ g. This is 72 g.

72 g of MO will contain 16 g of oxygen and hence 56 g of M, this being 1 mole of M. The key is **D**. You have to work very hard for the mark in this question!

27. Iodine starts at oxidation number −1 in the iodide ion and is converted to oxidation number +5 in the iodate ion, IO_3^-. The numerical change is 6 units, key **D**.

28. Sodium bromate and dilute acetic acid (ethanoic acid) do not react with bromine water. The other three substances do react with bromine water but use of two of these reagents would cause problems later in the experiment.

Sodium thiosulphate reacts with bromine water, one course of the reaction giving sulphate and bromide ions:

$$S_2O_3^{2-} + 4Br_2 + 5H_2O \rightarrow$$
$$2SO_4^{2-} + 8Br^- + 10H^+$$

The problem over using thiosulphate is that it is difficult to be certain that the exact amount required has been added. If too little is used, bromine will remain and will liberate extra iodine when the potassium iodide is added. If too much thiosulphate is used, it will react with some or all of the iodine liberated later and the final titration reading will be too low, possibly zero.

The silver ions in silver nitrate react with bromine water. The bromide ions in equilibrium with bromine molecules form a precipitate of silver bromide:

$$Br_2 (aq) + H_2O(l) \rightleftharpoons$$
$$Br^- (aq) + OBr^- (aq) + 2H^+ (aq)$$

The equilibrium position will be disturbed by removal of $Br^- (aq)$ and eventually all the bromine will be converted to insoluble AgBr and a solution of OBr^- ions (if sufficient silver nitrate is added). The OBr^- ions contain bromine of oxidation number +1 and, in the acidic solution that is present, this will oxidize iodide ions to iodine when potassium iodide is added later. The added iodide ions could also produce insoluble AgI if excess silver ions are present.

You may not have met yet the reaction of bromine water with phenol but, by elimination, it must be the correct choice of reagent. The reaction produces tribromophenol and ionized HBr:

$$C_6H_5OH + 3Br_2 \rightarrow$$
$$C_6H_2Br_3OH + 3H^+ + 3Br^-$$

Use of excess phenol will completely remove bromine molecules by this substitution reaction. Neither excess phenol nor any of the reaction products interfere with the course of the later stages of the determination so **E** is the correct key. This is a difficult problem!

29. The reaction between iodide ions and iodate ions to produce iodine is, in terms of oxidation numbers

$$I(-1) + I(+5) \rightarrow I(0)$$

Balancing 'ups and downs' in oxidation numbers gives

$$5I(-1) + I(+5) \rightarrow 6I(0)$$

Putting in the real nature of the oxidation numbers and balancing for atoms and charges with H_2O and H^+ gives

$$5I^- + IO_3^- + 6H^+ \rightarrow 3I_2 + 3H_2O$$

This shows that three moles of I_2 are formed from one mole of potassium iodate, assuming that excess iodide has been added. This is key **C**.

30. The titration required 12 cm^3 of 0·005M thiosulphate. This volume contains 0·005 × 12/1000 mole of thiosulphate. The number of moles of I_2 reacting with this thiosulphate is

$$\frac{1}{2} \times \frac{12}{1000} \times 0·005$$

Three moles of I_2 are formed from every mole of I^- present in the table salt. This is shown in the reaction scheme below.

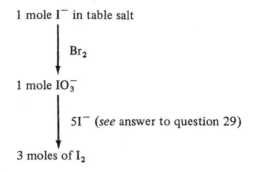

1 mole I^- in table salt

Br$_2$

1 mole IO_3^-

5I$^-$ (*see* answer to question 29)

3 moles of I_2

The number of moles of I^- originally present must be

$$\frac{1}{3} \times \frac{1}{2} \times \frac{12}{1000} \times 0·005 = \frac{0·01}{1000} = 10^{-5}$$

This is key **C**. Another difficult problem!

Test 5

Enthalpy changes

Questions 1–4 concern the following terms:

  A enthalpy of formation

  B enthalpy of neutralization

  C enthalpy of hydration

  D electron affinity

  E ionization energy

Select, from **A** to **E**, the term which best describes the energy change occurring in each reaction below.

1. $H(g) \rightarrow H^+(g) + e^-$

2. $H^+(g) + H_2O(l) \rightarrow H_3O^+(aq)$

3. $\frac{1}{2}H_2(g) \rightarrow H(g)$

4. $H_3O^+(aq) + OH^-(aq) \rightarrow 2H_2O(l)$

Directions summarized for questions 5 to 9				
A	**B**	**C**	**D**	**E**
1,2,3 only correct	1,3 only correct	2,4 only correct	4 only correct	Some other response or combination of responses is correct

5. In a simple experiment to determine the heat of combustion of an alcohol, a spirit burner containing the alcohol was weighed, lit and placed under an aluminium can containing water. The temperature rise of the water was determined and the spirit burner re-weighed. From the results the heat of combustion of the alcohol was determined. The results were numerically very much lower than the values obtained by accurate calorimetry. Which of the following factors would tend to make the result too low?

  1 Heat loss around the sides of the aluminium can during the experiment

  2 Evaporation of alcohol from the hot wick after extinguishing the flame and before re-weighing

  3 Evaporation of alcohol from the spirit lamp after the first weighing but before lighting

  4 The thermometer touching the bottom of the can

6. The equation for the combustion of octane is

$$C_8H_{18} + 12\tfrac{1}{2}O_2 \rightarrow 8CO_2 + 9H_2O;$$
$$\Delta H^{\ominus}_{298} = -5498 \text{ kJ}$$

   (C = 12, H = 1, O = 16)

   From this information alone, it can be deduced that

   1 if 114 g of octane are subjected to complete combustion, 5498 kJ will be released

   2 octane is a dangerously flammable liquid

   3 if 57 g of octane are burned, then 81 g of water will be formed

   4 if all volumes are measured at s.t.p., the reaction will be accompanied by an overall increase in volume

7. Which of the following experimentally determined quantities requires MORE THAN ONE thermochemical experiment?

   1 The lattice energy of calcium chloride

   2 The heat of formation of methane

   3 The energy required to atomize 1 mole of dimethyl ether, $(CH_3)_2O$

   4 The heat of combustion of decanol

8. Bond energy terms needed to calculate the approximate heat of atomization of 1-chloro-prop-l-ene include those for the bonds

   1 C–H

   2 C–C

   3 C–Cl

   4 C=C

9. In order to estimate the heat of formation of the hypothetical ionic compound 'MgCl', a number of pieces of data are required among which are the enthalpy changes of

1 $Cl(g) + e^- \rightarrow Cl^-(g)$

2 $Mg(s) \rightarrow Mg(g)$

3 $\tfrac{1}{2}Cl_2(g) \rightarrow Cl(g)$

4 $Mg^+(g) \rightarrow Mg^{2+}(g) + e^-$

Directions for questions 10 to 24. Each of the questions or incomplete statements in this section is followed by five suggested answers. Select the best answer in each case.

---

10. The standard enthalpies of formation for carbon dioxide and formic acid are -393.7 kJ mol^{-1} and -409.2 kJ mol^{-1} respectively. The enthalpy change, in kJ mol^{-1}, for the reaction

$$H_2(g) + CO_2(g) \rightarrow HCOOH(l)$$

   would be

   A -802.9

   B -414.7

   C -15.5

   D $+15.5$

   E $+802.9$

11.

	$\Delta H^{\ominus}_{f,\,298}$/kJ mol^{-1}
$CS_2(l)$	88
$NOCl(g)$	53
$CCl_4(l)$	-139
$SO_2(g)$	-296

   From the data above, the value of $\Delta H^{\ominus}_{298}$ in kJ for the reaction

$$CS_2(l) + 4NOCl(g) \rightarrow$$
$$CCl_4(l) + 2SO_2(g) + 2N_2(g)$$

   is

   A -1031

   B -731

   C -431

   D 431

   E 1031

12. The enthalpy changes for two reactions are given by the equations

$$2Cr(s) + 1\tfrac{1}{2}O_2(g) \rightarrow Cr_2O_3(s);$$
$$\Delta H = -1130 \text{ kJ}$$

$$C(s) + \tfrac{1}{2}O_2(g) \rightarrow CO(g); \qquad \Delta H = -110 \text{ kJ}$$

What is the enthalpy change, in kJ, for the reaction

$$3C(s) + Cr_2O_3(s) \rightarrow 2Cr(s) + 3CO(g)?$$

A −1460

B − 800

C + 800

D +1020

E +1460

13. The enthalpy of combustion of pentane, C_5H_{12}, is −3520 kJ mol^{-1}. The enthalpy of formation of CO_2 is −395 kJ mol^{-1} and that of H_2O is −286 kJ mol^{-1}.
The enthalpy of formation of pentane, in kJ mol^{-1}, is

A −7211

B −2839

C − 171

D + 171

E +2839

14. Which one of the following reactions is accompanied by an enthalpy change which is equal to the bond energy for H−I?

A $2HI(g) \rightarrow H_2(g) + I_2(g)$

B $HI(g) \rightarrow \tfrac{1}{2}H_2(g) + \tfrac{1}{2}I_2(g)$

C $HI(g) \rightarrow \tfrac{1}{2}H_2(g) + \tfrac{1}{2}I_2(s)$

D $HI(g) \rightarrow H(g) + I(g)$

E $HI(g) \rightarrow H^+(g) + I^-(g)$

15. Which of the following equations represents the change for which $\Delta H^{\ominus}_{298}$ would be equal to the lattice energy of calcium bromide?

A $Ca(g) + Br_2(g) \rightarrow CaBr_2(g)$

B $Ca^{2+}(s) + 2Br^-(s) \rightarrow CaBr_2(s)$

C $Ca^{2+}(g) + 2Br^-(g) \rightarrow CaBr_2(g)$

D $Ca^{2+}(aq) + 2Br^-(aq) \rightarrow CaBr_2(s)$

E $Ca^{2+}(g) + 2Br^-(g) \rightarrow CaBr_2(s)$

16. Which change would have a negative ΔH value?

A $Na(s) \rightarrow Na(g)$

B $Na(g) \rightarrow Na^+(g) + e^-$

C $Na^+Cl^-(s) \rightarrow Na^+(g) + Cl^-(g)$

D $Cl_2(g) \rightarrow 2Cl(g)$

E $Cl(g) + e^- \rightarrow Cl^-(g)$

17. In the energy cycle given below, what is the value (in kJ mol^{-1}) for the electron affinity of bromine?

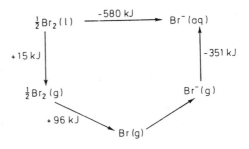

A −820

B −340

C −118

D +118

E +340

18. In which of the following would you expect the enthalpy change to be greatest?

A $CH_2 = CH-CH_3(g) \rightarrow CH_2 - CH_2(g)$
               $\quad\quad\quad\quad\quad\quad\quad \searrow CH_2 \swarrow$

B $F_2(s) \rightarrow F_2(l)$

C $Mg^{2+}(g) + O^{2-}(g) \rightarrow MgO(s)$

D $Br(g) + e^- \rightarrow Br^-(g)$

E $Cs(g) \rightarrow Cs^+(g) + e^-$

19. Some heats of formation, in kJ mol^{-1}, at 298 K are

S_8(rhombic)	0
S_8(g)	103
S(g)	274

What is the heat of atomization of sulphur in kJ mol^{-1}?

A $\dfrac{103}{8}$

B 103

C 274

D $274 + \dfrac{103}{8}$

E $274 + 103$

20. The following are bond energy terms:

C–C 348 kJ mol^{-1}
C–H 413 kJ mol^{-1}
C–O 357 kJ mol^{-1}

The enthalpy of formation, in kJ mol^{-1}, of gaseous dimethyl ether (CH_3OCH_3) from gaseous atoms is

A +3192

B +2835

C −2835

D −3183

E −3192

21. Some bond energy terms at 298 K in kJ mol^{-1} are:

C–C	347
C=C	613
C–H	416
H–H	437

Which of the following is $\Delta H^{\ominus}_{298}$ (in kJ mol^{-1}) for the reaction represented by the equation

$CH_2 = CHCH_2CH_3(g) + H_2(g) \rightarrow$
$\quad\quad\quad\quad\quad\quad CH_3CH_2CH_2CH_3(g)?$

A −395

B −129

C +129

D +287

E +395

22. The enthalpy of atomization of graphite is 725 kJ mol^{-1} and that of hydrogen is 218 kJ per mole of hydrogen atoms. The enthalpy of formation of methane is −76 kJ mol^{-1}.

The enthalpy of formation of the C–H bond from gaseous carbon and atomic hydrogen is

A +418 kJ

B +255 kJ

C +76 kJ

D −255 kJ

E −418 kJ

23. For the reaction

$$H_2(g) + \tfrac{1}{2}O_2(g) \rightarrow H_2O(g);$$
$$\Delta H^{\ominus}_{298} = -242 \text{ kJ}$$

the enthalpies of dissociation of hydrogen and oxygen are given by

$$H_2(g) \rightarrow 2H(g); \quad \Delta H^{\ominus}_{298} = +436 \text{ kJ mol}^{-1}$$
$$O_2(g) \rightarrow 2O(g); \quad \Delta H^{\ominus}_{298} = +500 \text{ kJ mol}^{-1}$$

The bond energy of the O—H bond in water is

A +121 kJ

B +242 kJ

C +444 kJ

D +464 kJ

E +589 kJ

24. Given the data

$$H_3O^+(aq) + OH^-(aq) \rightarrow 2H_2O(l);$$
$$\Delta H^{\ominus}_{298} = -57.2 \text{ kJ mol}^{-1}$$

$$CH_3CO_2H(aq) + OH^-(aq) \rightarrow$$
$$CH_3CO_2^-(aq) + H_2O(l);$$
$$\Delta H^{\ominus}_{298} = -56.0 \text{ kJ mol}^{-1}$$

calculate the enthalpy change (in kJ mol^{-1}) for the following reaction, assuming it goes to completion:

$$CH_3CO_2H(aq) + H_2O(l) \rightarrow$$
$$CH_3CO_2^-(aq) + H_3O^+(aq)$$

A +113·2   D − 56·6

B + 1·2   E −113·2

C − 1·2

25. Given the following information, what is the enthalpy of formation of solid phosphorus pentachloride in KJ mol^{-1}?

$$PCl_3(l) + Cl_2(g) \rightarrow PCl_5(s);$$
$$\Delta H = -137 \text{ kJ mol}^{-1}$$
$$P_4(s) + 6Cl_2(g) \rightarrow 4PCl_3(l);$$
$$\Delta H = -1264 \text{ kJ mol}^{-1}$$

A −179   D −1401

B −453   E −1812

C −1127

26. Given the following data :

	ΔH
$\tfrac{1}{2}N_2(g) + \tfrac{3}{2}H_2(g) \rightarrow NH_3(g)$	−46 kJ
$\tfrac{1}{2}H_2(g) \rightarrow H(g)$	+218 kJ
$\tfrac{1}{2}N_2(g) \rightarrow N(g)$	+473 kJ

the mean energy (in kJ mol^{-1}) required to break the N—H bond in NH$_3$ is

A +1173   D −391

B +391   E −1173

C +215

27. The following is a list of bond energy terms, in kJ mol^{-1} .

C—C	348
C=C	612
Cl—Cl	242
C—Cl	338

Which one of the following expressions will give the correct value for the enthalpy change (in kJ mol^{-1}) of the addition reaction between chlorine and ethene?

A 612 + 242 + 348 + (2 × 338)

B −612 −242 − 348 − (2 × 338)

C −612 − 242 + 348 + (2 × 338)

D 612 + 242 − 348 − (2 × 338)

E 612 + 242 + 348 − (2 × 338)

Questions 28–31 concern the following diagram which represents an energy cycle for the formation of potassium chloride.

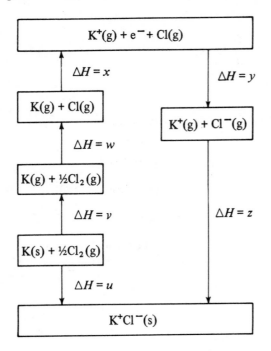

Select the best answer for each question below.

28. Which one of the following is $\Delta H_f^{\ominus}$ for potassium chloride?

A u

B z

C $z - u$

D $x + y + z$

E $u + w$

29. Which one of the following represents the lattice energy for potassium chloride?

A $z - u$

B u

C $u + w$

D $x + y$

E z

30. Which of the following represents the electron affinity for chlorine?

A w

B y

C $y + w$

D $x + y$

E $z + w$

31. Which of the following is a correct statement about x?

A It is the lattice energy for potassium.

B It is the energy required to vaporize 1 mole of potassium.

C It is the heat of atomization for potassium.

D It is the first ionization energy for potassium.

E It is the heat of formation for potassium.

Test 5   Answers

1. The loss of an electron from a gaseous atom is the first ionization energy. This is key **E**. Does hydrogen have a second ionization energy?

2. The energy change when a gaseous ion is dissolved in water is the enthalpy of hydration of that ion, key **C**. It is possible that you may not have met this term at the present stage in your course.

3. The energy change when a substance, in this case gaseous hydrogen atoms, is formed from its elements in their standard state, in this case hydrogen gas, is called the enthalpy of formation. The key is **A**. The quantity in this example is also half the bond dissociation energy of the hydrogen molecule.

4. The reaction is neutralization and the key is **B**.

5. Heat loss around the sides of the can is the major reason for the method giving very low values. The first statement is correct. It is possible to improve the value by modifying the procedure and equipment so that more energy goes into the water and less into the room. The only way, however, to obtain a value by this method that is anywhere near the data book value is to first calibrate the apparatus for heat loss. The experiment is first carried out using an alcohol of known heat of combustion and the fraction of the energy lost to the room is determined. The experiment is repeated with the alcohol of unknown heat of combustion. The conditions are kept as near identical as possible to those in the calibration experiment and the already calculated correction is made for heat loss.

   Both **2** and **3** lose alcohol between weighings but not by combustion. This will help to reduce the experimental value for the heat of combustion.

   The thermometer touching the bottom of the can might cause the temperature recorded to be too high and certainly will do so if the water is inefficiently stirred. The temperature rise recorded might be too high (it certainly will not be too low) and this would make the calculated heat of combustion too high. This error would never, however, compensate for the large error discussed in **1**. Only **1, 2** and **3** are correct so the key is **A**.

6. 1 mole of octane has a mass of 114 g and the equation contains the information that 5498 kJ will be released on complete combustion of this mass. The first statement can be deduced from the information alone.

   2 is correct but cannot be deduced from the information given.

   The equation gives information about moles and shows that if 0·5 mole of octane (57 g) is burned, 4·5 mole of water (81 g) will be formed.

   The equation does not include state symbols so that nothing can be deduced about volume changes from the information given. Using other information, only oxygen and carbon dioxide are gases at s.t.p., the other two substances are liquids. There will be an overall volume decrease as 12·5 volumes become 8 volumes. This is replacing moles by volumes, the reverse of test 2, question 5. Even if the fourth statement could be deduced, it is incorrect! Only **1** and **3** can be deduced from the information so the key is **B**.

7. Only **4**, the heat of combustion, can be found by one thermochemical experiment. If the quantities listed in **1, 2** and **3** are being determined by thermochemical methods, each requires more than one experiment. The correct key is **A**.

   The lattice energy of $CaCl_2$ could be determined theoretically by one lengthy calculation of all the energy changes involved, on attraction and repulsion, as one mole of gaseous Ca^{2+} ions and two moles of gaseous Cl^- ions, all at infinite separation, are gradually assembled in the $CaCl_2$ lattice. Experimentally, however, it would require determination of all the other quantities in the Born–Haber cycle to find the lattice energy of $CaCl_2$. Can you draw an energy level diagram showing all the energy changes involved? There is a similar diagram for KCl in questions 28 to 31.

It is not possible to bring carbon and hydrogen together in a calorimeter and persuade them to form methane. Carbon and hydrogen do not react except at very high temperatures when small quantities of ethyne, C_2H_2, rather than methane, are formed. Can you suggest one reason for the lack of reaction between carbon and hydrogen? The heat of formation of methane requires experimental determination of the separate heats of combustion of methane, carbon and hydrogen. Can you put these in a suitable energy level diagram? Question 13 deals with a similar problem for pentane.

The enthalpy of atomization of dimethyl ether could be found by spectroscopic studies on the gradual, complete fragmentation of the molecule. It could also be found by determining the following, in which (1) and (2) are thermochemical experiments:

(1) the heat of combustion of dimethyl ether in a bomb calorimeter;
(2) the heats of combustion of carbon and hydrogen in a bomb calorimeter;
(3) the enthalpy of atomization of carbon from vapour pressure measurements and the bond dissociation energies of hydrogen and oxygen from spectroscopic studies.

Can you put (1), (2) and (3) in a suitable energy level diagram which also includes the enthalpy of atomization of dimethyl ether?

8. The compound has the structure

It contains all the bonds given in the question, the bond energy terms of all of which are required to calculate the heat of atomization of the compound. The correct key is **E**. Why is the word 'approximate' used in the question?

9. A Born–Haber cycle to determine the heat (enthalpy) of formation of 'MgCl' would require **1**, **2** and **3** but not **4**. This last energy change would be needed for a Born–Haber cycle on $MgCl_2$. The correct key is **A**.

What other quantities apart from **1**, **2** and **3** are needed in order to be able to estimate the heat of formation of 'MgCl'? You could refer to the energy level diagram in questions 28–31 to check your answer.

10. The standard enthalpy of formation of hydrogen gas is zero as it is an element in its standard state. Putting values for $\Delta H_f^{\ominus}$ below the equation gives

$$H_2(g) + CO_2(g) \rightarrow HCOOH(l)$$
$$\Delta H_f^{\ominus}/kJ\ mol^{-1}\quad 0\qquad -393.7\qquad -409.2$$

Compared with the standard state, the formation of products is more exothermic than the formation of reactants. The enthalpy change in the reaction is $-15.5\ kJ\ mol^{-1}$, key **C**.

This could also be shown on an energy level diagram, with energies in $kJ\ mol^{-1}$

11. The standard enthalpy of formation of nitrogen is zero—element in the standard state. Putting values below the equation gives

$$CS_2(l) + 4NOCl(g) \rightarrow$$
$$\Delta H_{f,298}^{\ominus}/kJ\ mol^{-1}\ \underbrace{+88\qquad 4 \times +53}_{+300}$$

$$CCl_4(l) + 2SO_2(g) + 2N_2(g)$$
$$\underbrace{-139\qquad\qquad 2 \times -196\quad 2 \times 0}_{-731}$$

Compared with the standard state, the formation of products is more exothermic than the

formation of reactants by 1031 kJ mol^{-1}.
$\Delta H_{298}^{\ominus}$ for the reaction is -1031 kJ mol^{-1},
key **A**.

12. The first equation gives the enthalpy of formation of Cr_2O_3, the second that of CO. Assuming
that these are standard values, they can be
put below the equation with C and Cr set at
zero.

$$3C(s) + Cr_2O_3(s) \rightarrow$$
$$\Delta H_f^{\ominus}/\text{kJ mol}^{-1} \quad 0 \qquad -1130$$

$$2Cr(s) + 3CO(g)$$
$$0 \qquad 3 \times -110$$

Compared with the standard state, the formation of products is less exothermic than the
formation of reactants by 800 kJ mol^{-1}. The
enthalpy change in the reaction is $+800$ kJ mol^{-1},
key **C**.

13. The equation for the combustion of pentane is
written, with enthalpies of formation below the
formulae

$$C_5H_{12}(l) + 8O_2(g) \rightarrow$$
$$\Delta H_f^{\ominus}/\text{kJ mol}^{-1} \quad p \qquad 0$$

$$5CO_2(g) + 6H_2O(l)$$
$$5 \times -395 \qquad 6 \times -286$$
$$\underbrace{\qquad\qquad}_{-3691}$$
$$\Delta H^{\ominus} = -3520 \text{ kJ mol}^{-1}$$

The enthalpy change in this reaction is
$-3691 - p$ and is also known to be -3520.
p is -171, key **C**.

14. The bond energy of HI is the energy change
when one mole of gaseous HI is converted to
gaseous atoms of H and I. This is key **D**. In a
question of this type, you must look carefully
at the state symbols before deciding on the
correct answer. This is also the case in question
15.

15. The lattice energy of $CaBr_2$ is the energy change
when one mole of $CaBr_2(s)$ lattice is formed
from the necessary number of gaseous ions of
Ca^{2+} and Br^- at infinite separation. This is
represented by equation **E**.

16. A negative ΔH value shows that one is looking
for an exothermic reaction. The sublimation of
sodium (**A**), ionization of gaseous sodium (**B**),
break-up of the sodium chloride lattice (**C**) and
bond breaking in Cl_2 (**D**) are all endothermic.
The change in **E** involves the electron affinity
of chlorine. This is an exothermic reaction and
the correct key is **E**.
    Check in a data book the two values for the
electron affinity of oxygen. Why does oxygen
have two values? Try to write equations for
the two reactions involved and explain why the
ΔH values have different signs.

17. The electron affinity of bromine is the energy
change in the diagram for which no value is
given.

$$Br(g) + e^- \rightarrow Br^-(g); \qquad \Delta H = x$$

The values in the diagram can be used to find
two values for the energy change on going from
$\frac{1}{2}Br_2(l)$ to $Br^-(aq)$.

  For the direct route
    $\Delta H = -580$ kJ

  For the indirect route
    $\Delta H = +15 + 96 + x - 351$ kJ

The energy change is independent of the route
so these two ΔH values are the same and x is
-340, key **B**.

18. The enthalpy change on rearrangement and
melting, **A** and **B**, will be small. The choice is
between **C**, **D** and **E**.
    D is an electron affinity — if you have
answered question 17 correctly, you will know
what value this quantity has for bromine.
Electron affinities of this type do not have very
high values, nor do first ionization energies for
alkali metals, particularly if the element is of
high atomic number. Ionization energies can
have very large values, especially if a new
electron shell is involved. **E** would be a better
candidate for the correct key if it was the
second ionization energy of caesium.
    C is a lattice energy. For a 1+ metal ion and
a 1− non-metal ion the lattice energy is of the
order of 800 kJ mol^{-1}. In **C**, however, both ions
carry a double charge and, because of the extra
attraction involved, this will give an approximate

two-fold doubling of the lattice energy; the value will be about four times that of a 1+, 1− lattice and this makes **C** the correct key.

Which key would you select if **C** and **E** were:

C $Na^+(g) + Cl^-(g) \rightarrow NaCl(s)$

E $Cs^+(g) \rightarrow Cs^{2+}(g) + e^-$

19. The heat of atomization of sulphur is the energy required to produce one mole of mon-atomic sulphur gas at 298 K from sulphur in the standard state. This is for the reaction

$$\tfrac{1}{8}S_8(s) \rightarrow S(g)$$

$\Delta H_f^{\ominus}$/kJ mol^{-1} 0   +274

ΔH for the reaction is +274, key **C**.

20. You will probably be more familiar with bond breaking than with bond forming:

H—C—O—C—H → 6H(g) + 2C(g) + O(g)

The conversion of dimethyl ether into separate gaseous atoms requires the breaking of six C—H bonds and two C—O bonds. The energy change will be $6 \times 413 + 2 \times 357$ which is +3192 kJ mol^{-1}. The question is about the reverse of this process so the energy change will be −3192 kJ mol^{-1}, key **E**.

21. The bond energy terms can be used to find enthalpy changes in reactions by linking reactants and products to the same constant energy level, that of the separate gaseous atoms. It is similar to the more usual method for predicting enthalpy changes by linking reactants and products to another constant level, that of elements in their standard states.

The breakdown of the reactants into gaseous atoms is

C=C C—H + H₂ → gaseous atoms

Energy change =		kJ mol^{-1}
one C=C bond		613
two C—C bonds		694
eight C—H bonds		3328
one H—H bond		437
Total		5072

The breakdown of the product into gaseous atoms is

H—C—C—C—C—H → gaseous atoms

Energy change =		kJ mol^{-1}
three C—C bonds		1041
ten C—H bonds		4160
Total		5201

It requires 129 kJ mol^{-1} more to break the bonds in the products than in the reactants so ΔH for the reaction in the problem is −129 kJ mol^{-1}, key **B**.

The molecules and gaseous atoms could be put on an energy level diagram

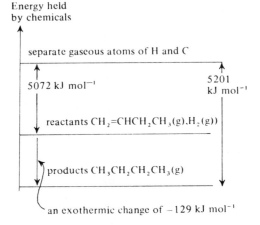

You may have noticed that in the route chosen to go from reactants to product, most of the bonds that are broken are then formed again to produce the product. Another, simpler route, which is probably close to that taken by the real reaction, is as follows.

Break one bond of the double bond.
 This requires $613 - 347$ kJ mol^{-1}.
Dissociate the hydrogen molecules into atoms.
 This requires 437 kJ mol^{-1}.
Form two new C–H bonds per double bond broken.
 This evolves 2×416 kJ mol^{-1}.

The total energy change is $(613 - 347) + 437 - 2 \times 416 = -129$ kJ mol^{-1}.

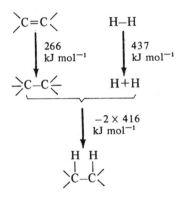

22. The values in the question can be put in an energy level diagram.

Energy held
by chemicals

separate gaseous atoms	C(g), 4H(g)	
	725 kJ $\quad$ 4 × 218 kJ	
standard state	C(s), $\quad$ 2H$_2$(g)	
	76 kJ	
	CH$_4$(g)	

The enthalpy of formation of methane from separate gaseous atoms is the change from the top to the bottom level in the diagram. This is $-[725 + (4 \times 218) + 76]$ or -1673 kJ mol^{-1}. The enthalpy of formation of one mole of C–H bonds is one quarter of this value or -418 kJ mol^{-1}, key **E**.

23. It is necessary to construct an energy level diagram similar to that in question 22.

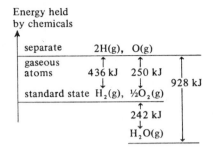

The energy change in the reaction

$$H_2O(g) \rightarrow 2H(g) + O(g)$$

is $(242 + 436 + 250)$ kJ mol^{-1} or 928 kJ mol^{-1}. This reaction involves the breaking of two moles of O–H bonds so the mean bond energy is half this value, 464 kJ mol^{-1}, which is key **D**.

24. Consider the two reactions at the start of the question. If the reverse of the first (the ionization of water) is followed by the second, the overall effect is to give the reaction for which the energy change is required.

$\Delta H_{298}^{\ominus}$
kJ mol^{-1}

$H_2O(l) + H_2O(l) \rightleftharpoons$
$\quad H_3O^+(aq) + OH^-(aq) \qquad +57\cdot2$
then
$CH_3CO_2H(aq) + OH^-(aq) \rightleftharpoons$
$\quad CH_3CO_2^-(aq) \quad + H_2O(l) \qquad -56\cdot0$
adding gives
$CH_3CO_2H(aq) + H_2O(l) \rightleftharpoons$
$\quad CH_3CO_2^-(aq) \quad + H_3O^+(aq) \qquad + 1\cdot2$

The correct key is **B**. Note that all three reactions are acid–base and are linked by an equilibrium sign. Can you identify acid 1, base 1, acid 2 and base 2 in each of these reactions?

25. One mole of solid phosphorus pentachloride can be formed from its elements in their standard states (solid phosphorus and gaseous chlorine) in two steps. The first step is the formation of one mole of $PCl_3(1)$. The enthalpy change for this step is $-1264/4$ kJ mol^{-1} which is -316 kJ mol^{-1}. The second step is the conversion of one mole of $PCl_3(1)$ to one mole of $PCl_5(s)$. The enthalpy change in the second step is -137 kJ mol^{-1}. The combined enthalpy change for the two steps is -453 kJ mol^{-1} and this is the value given in key **B**.

26. One mole of gaseous ammonia can be converted to gaseous molecules of nitrogen and hydrogen by two different routes. The enthalpy changes in these routes can be calculated from the information given in the question.

One route is the one-step reversal of the reaction represented by the first equation in the question. The enthalpy change by this route is $+46$ kJ mol^{-1}. Note that the sign has been reversed.

The first step in the other route involves breaking the three N–H bonds in every molecule in one mole of ammonia. If the mean bond energy of N–H is x, the enthalpy change in this step is $3x$. The first step produces one mole of gaseous nitrogen atoms and three moles of gaseous hydrogen atoms. The second step is the combination of these gaseous atoms to form gaseous molecules of nitrogen and hydrogen. The total enthalpy change in the second step is $-473 -(3 \times 218)$ kJ mol^{-1}.

The enthalpy changes by the two different routes must be the same, so

$$+46 \ = \ 3x - 473 - (3 \times 218)$$

$$3x \ = \ +1173$$

$$x \ = \ +391$$

This shows that the key is **B**.

If you have difficulty with this question, try to draw an energy level diagram similar to the diagrams given in the answers to questions 21, 22 and 23. The three levels you need on your diagram are :

1. $\frac{1}{2}N_2(g)$ and $1\frac{1}{2}H_2(g)$

2. $N(g)$ and $3H(g)$

3. $NH_3(g)$

27. The addition reaction between chlorine and ethene could take place by a two-step route involving the breaking or forming of the bonds whose bond energies are given in the question. The first step would be to break the $C = C$ bond and the $Cl - Cl$ bond, leaving all $C - H$ bonds intact. The enthalpy change in this first step is $612 + 242$ kJ mol^{-1}.

The second step would be to form the $C - C$ bond and two $C - Cl$ bonds. The enthalpy change in this step is $-348 -(2 \times 338)$ kJ mol^{-1}.

The enthalpy change in the complete reaction is the sum of the values for the two steps. This sum is

$$612 + 242 - 348 - (2 \times 338).$$

This is shown in key **D**.

28. The enthalpy of formation of KCl, $\Delta H_f^{\ominus}$, is the energy change when it is formed from its elements in their standard states. This is u, key **A**.

29. The lattice energy of KCl is the energy change when the $K^+Cl^-(s)$ lattice is formed from its gaseous ions. This is z, key **E**.

30. The electron affinity of chlorine is the energy change when gaseous Cl^- ions are formed from gaseous Cl atoms. This is y, key **B**.

31. The change for which $\Delta H = x$ involves the conversion of gaseous potassium atoms into gaseous potassium ions by electron loss. This is the first ionization energy of potassium, key **D**.

Test 6

Structure and bonding

Questions 1–4 concern the following shapes of particles:

    A linear

    B square planar

    C trigonal planar

    D trigonal pyramidal

    E tetrahedral

Which best describes the shape of each of the following ions or molecules?

1. BCl_3

2. Ammonium ion, NH_4^+

3. Ammonia, NH_3

4. Nitrate ion, NO_3^-

Questions 5–7 concern the following techniques:

    A atomic emission spectroscopy

    B infrared spectroscopy

    C *X*-ray crystallography

    D mass spectrometry

    E electron diffraction

Select from **A** to **E** the technique which will

5. determine the N–H stretching frequency in aniline

6. determine the bond lengths and bond angles in a gaseous molecule by interaction with a stream of particles

7. determine the palladium–gold interatomic distance in an alloy of palladium and gold

Directions summarized for questions 8 to 14				
A	**B**	**C**	**D**	**E**
1,2,3 only correct	1,3 only correct	2,4 only correct	4 only correct	Some other response or combination of responses is correct

8. Which of the following molecules would be expected to be linear?

    1 H_2S

    2 PH_3

    3 H_2O

    4 CO_2

9. Which of the following, in the solid state, has a crystal structure which contains discrete molecules?

    1 Magnesium oxide

    2 Carbon dioxide

    3 Silicon dioxide

    4 Rhombic sulphur

10. Which of the following will apply to the structure of sodium chloride?

    1 It is an example of hexagonal close packing.

    2 The distance between the nuclei of adjacent ions of opposite charge equals the sum of the ionic radii.

    3 It is the same as the structure of caesium chloride.

    4 Each Na^+ ion is surrounded by six Cl^- ions.

11. Which of the following can accept an electron pair in the formation of a dative covalent bond?

    1 NH_3

    2 $AlCl_3$

    3 CH_4

    4 BF_3

12. Delocalization of electrons may be expected in the species of formula

    1 N_2H_4

    2 NO_3^-

    3 HNO_3

    4 NH_4^+

13.

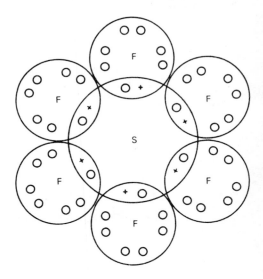

The diagram represents the electronic structure of the sulphur hexafluoride molecule.

Correct statements about sulphur hexafluoride include that

1 all S–F bonds are equivalent

2 SF_6 is a planar molecule

3 the oxidation number of sulphur is the same as the number of electrons it uses in bonding

4 sulphur has acquired the electronic structure of the inert gas argon

14. The infrared absorption spectra of decane, trichloromethane and tetrachloromethane are shown below.

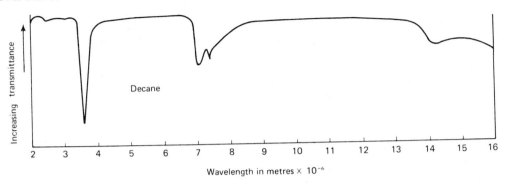

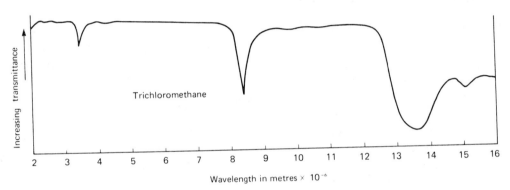

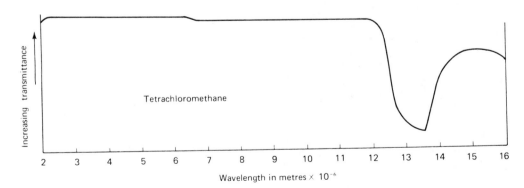

Reasonable deductions from these spectra include that

1 C—H bonds absorb radiation of wavelength 3.5×10^{-6} metres

2 C—C bonds absorb radiation of wavelength 7×10^{-6} metres

3 C—Cl bonds absorb radiation of wavelength 13.5×10^{-6} metres

4 C—H bonds absorb radiation of wavelength 8.3×10^{-6} metres

Directions for questions 15 to 20. Each of the
questions or incomplete statements in this section
is followed by five suggested answers. Select the
best answer in each case.

15. Hydrogen atoms are not usually detected by
X-ray diffraction. The reason is that the

   A mass of the hydrogen atom is too small to
      deflect X-rays appreciably

   B electron density in the hydrogen atom is
      too low to cause diffraction

   C radius of a hydrogen atom is less than the
      wavelength of the X-rays used

   D hydrogen nucleus does not absorb radiation
      in the X-ray region

   E bonds to hydrogen atoms are so short that
      the hydrogen atoms are masked by other
      atoms

16. The diagram below is of the unit cell of the
   'perovskite' lattice.

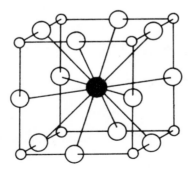

   Europium, titanium and oxygen form a compound having this structure.

   The formula of the compound is therefore

   A $EuTiO_3$

   B $EuTi_2O_3$

   C $EuTi_8O_{12}$

   D Eu_2TiO_2

   E Eu_2TiO_6

17. The distance between adjacent, oppositely
   charged ions in rubidium chloride is 3·285Å;
   in potassium chloride is 3·139 Å; in sodium
   bromide is 2·981 Å and in potassium bromide
   is 3·293 Å (1 Å = 10 nm).

   The distance between adjacent oppositely
   charged in ions in rubidium bromide is

   A 3·147 Å

   B 3·385 Å

   C 3·393 Å

   D 3·439 Å

   E 3·605 Å

18. In which one of the following molecules woul
   you expect to find the smallest angle between
   two adjacent covalent bonds?

   A BeH_2

   B BF_3

   C CCl_4

   D NH_3

   E OH_2

19. The data below refer to eight elements, lettered *M* to *T* (these letters are NOT chemical symbols).

Element	*M*	*N*	*O*	*P*	*Q*	*R*	*S*	*T*
Atomic number	Z	Z + 1	Z + 2	Z + 3	Z + 4	Z + 5	Z + 6	Z + 7
Molar enthalpy of vaporization (kJ mol^{-1})	2.8	3.4	3.3	1.8	89	129	294	377
Boiling point (K)	73	93	83	23	1163	1373	2673	2973

From these data, it can be deduced that

A *T* is in the same Group of the Periodic Table as helium

B *S* has a giant structure

C *N* is a metallic element

D the elements are all in the same period of the Periodic Table

E *M* is a Group 1 element in the Periodic Table

20. The electronegativities of the elements *P, Q, R, S* and *T* are given below:

P 0·7

Q 1·1

R 1·6

S 2·5

T 1·7

P, Q, R, S and *T* are NOT the chemical symbols for the elements.

Which of the following bonds has the most ionic character?

A *P–T* **D** *T–S*

B *P–Q* **E** *Q–T*

C *R–S*

21. The lattice energy of magnesium chloride is numerically equal to the energy change for the reaction

A $Mg^{2+}(s) + 2Cl^{-}(s) \rightarrow MgCl_2(s)$

B $Mg^{2+}(g) + 2Cl^{-}(g) \rightarrow MgCl_2(g)$

C $Mg(g) + Cl_2(g) \rightarrow MgCl_2(g)$

D $Mg^{2+}(g) + 2Cl^{-}(g) \rightarrow MgCl_2(s)$

E $Mg^{2+}(aq) + 2Cl^{-}(aq) \rightarrow MgCl_2(s)$

22. For which equation is the energy change called the bond energy term of the C—H bond?

A $¼CH_4(g) \rightarrow ¼C(g) + H(g)$

B $CH_4(g) \rightarrow C(s) + 2H_2(g)$

C $CH_4(g) \rightarrow C(g) + 4H(g)$

D $¼CH_4(g) \rightarrow ¼C(s) + ½H_2(g)$

E $CH_4(g) \rightarrow C(g) + 2H_2(g)$

23. The $C = C$ double bond has a bond length of 0.134 nm and a bond energy term of 610 kJ mol^{-1}. Which of the following pairs of figures is most likely to be correct for a $C - C$ single bond?

	Bond length /nm	Bond energy term /kJ mol^{-1}
A	0.067	305
B	0.124	585
C	0.134	625
D	0.154	345
E	0.154	835

24. Which of the following analytical techniques makes use of vibration within molecules?

    A Electron diffraction

    B Atomic emission spectroscopy

    C Infra-red absorption spectroscopy

    D Mass spectroscopy

    E X-ray diffraction

25. Which structure has an unpaired electron?

    A $N = O$

    B $H—O—N \overset{\nearrow O}{\underset{\searrow O}{}}$

    C NH_3

    D $N \equiv N$

    E $H—C \equiv N$

26. Which of the following does NOT have a planar structure?

    A NO_3^-

    B H_2CO

    C NH_3

    D $H_2C = CH_2$

    E BF_3

27. Which of the following species has the same total number of electrons as a chlorine atom?

    A Cl^+

    B HCl

    C Ne^-

    D OH^-

    E S^-

28. Which of the following has the smallest ionic radius?

    A Mg^{2+} D Na^+

    B F^- E O^{2-}

    C Cl^-

Questions 29–31

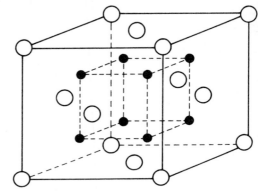

The diagram represents a unit cell of $SrCl_2$. The open circles, representing Sr^{2+} ions, are at the corners, and at the intersections of the face diagonals, of the big cube. The black circles, representing Cl^- ions, are on the body diagonals of the big cube, lying at the corners of the small cube shown by the dotted lines.

29. Which of the following statements is true of the structure of strontium chloride?

    A The structure is of the same type as that of sodium chloride.

    B The strontium ions are in a body-centred cubic arrangement.

    C The strontium ions are in a face-centred cubic arrangement.

    D Each chloride ion is at the centre of a cube of 8 strontium ions.

    E Each strontium ion is at the centre of a tetrahedron of 4 chloride ions.

30. The co-ordination numbers of Sr^{2+} and Cl^- ion respectively are

    A   4 : 8

    B   8 : 4

    C   8 : 8

    D   8 : 12

    E   12 : 8

31. If the volume of 1 mole of strontium chloride is V cm^3 and the volume of a unit cell is v cm^3, the Avogadro constant, L, is given by

    A $V/v \times \frac{1}{4}$

    B $V/v \times \frac{1}{2}$

    C V/v

    D $V/v \times 4$

    E $V/v \times 12$

Test 6   Answers

1 to 4.

When deciding the shape that a covalently bonded molecule is likely to take up, it is necessary to consider two points:

(1) the number of atoms that are arranged around any central atom in which you are interested;

(2) whether the central atom has any lone pairs of electrons that it is not using in bonding.

The pairs of electrons in bonds and any remaining lone pairs can be regarded as deciding the shape of the molecule by repulsion. The arrangement of the atoms attached to the central atom is one where repulsion between electron pairs has been minimized. The electron pairs are as far from each other as possible and this frequently makes the molecule take up a three-dimensional rather than a two-dimensional shape.

In the examples given in the Chart, each bond joining atoms represents a shared electron pair. Each atom supplies one electron in normal covalent bonds, while in dative covalent bonds both electrons are supplied by one atom. Any remaining electron pairs on the central atom are also shown. The bond angles given are the values when all the groups around the central atom are identical.

The shapes of the molecules in questions **1** to **4** can now be checked.

1 BCl_3 is trigonal planar, key **C**.

2 The NH_4^+ ion is tetrahedral, key **E**.

3 NH_3 is trigonal pyramidal, key **D**. Note that the electron pairs, in bonds and lone pairs, have a tetrahedral arrangement but the arrangement of atoms of H around N is not tetrahedral.

4 The nitrate ion is trigonal planar, key **C**.

Keys **A** and **B** have not been used. Can you find a molecule in the examples given earlier which could fit each of these keys? **B** is an unusual shape and you may have difficulty finding it. Look for a molecule where four atoms could be forced into a square planar arrangement around the central atom by the repulsion of two 'invisible' lone pairs.

5. Infrared spectroscopy is the technique used to determine stretching frequencies of bonds in molecules. This is key **B**. Examples of infrared spectra are given in question 14.

6. The technique referred to here is electron diffraction, key **E**. The other important approach to obtaining information about the shapes of gaseous molecules involves spectroscopic studies: infrared, microwave and Raman.

7. The technique required here must be capable of locating the atoms in a solid. The only possibility among those listed is X-ray crystallography, key **C**. For this to be successful, the solid must have a regular structure. The existence of such a structure in the alloy is implied by the wording of the question which suggests that a palladium–gold interatomic distance is waiting to be determined.

8. Refer to the diagrams in the answer to questions 1 to 4. CO_2 is the only linear molecule, key **D**.

CHART

Two atoms around each central atom. Bond angles 180 degrees. Shape is called 'linear'

H——Be——H   O══C══O   H—C≡≡C—H

Three atoms around each central atom. Bond angles 120 degrees. Shape is called 'trigonal planar', 'triangular planar' or 'planar triangular'

Two atoms and one lone pair around central atom. Bond angle 120 degrees. Shape is called 'bent' or '**V** shaped'

Four atoms around central atom. Bond angles close to 109 degrees. Shape is called 'tetrahedral'

Three atoms and one lone pair around central atom. Bond angles close to 109 degrees. Shape is called 'trigonal pyramidal' or 'pyramidal'

Two atoms and two lone pairs around central atom. Bond angles close to 109 degrees. Shape is called 'bent' or '**V** shaped'

Five atoms around central atom. Bond angles are 120 degrees and 90 degrees. Shape is called 'trigonal bipyramidal'. Similar shapes are taken up by four atoms and one lone pair, SF_4, and by three atoms and two lone pairs, ICl_3.

Six atoms around central atom. Bond angles are all 90 degrees. Shape is called 'octahedral'. Similar shapes are taken up by five atoms and one lone pair, IF_5, and four atoms and two lone pairs, XeF_4.

9. Discrete molecules are only possible in non-metallic elements and compounds between non-metals. Compounds between metals and non-metals, such as **1**, contain ions.

Non-metallic elements and compounds between them can be composed of small molecules or the covalent bonding can extend indefinitely to give a giant structure, as happens in silicon dioxide. Carbon dioxide and rhombic sulphur contain discrete molecules, CO_2 and S_8. The key is **C**.

10. **1** is incorrect. In sodium chloride, the Na^+ ions are in the face-centred cubic or cubic close packing arrangement. Note that the Na^+ ions do not touch so the arrangement is not described as cubic closest packing which is the arrangement of the atoms of many metallic elements. The Cl^- ions are also in their own face-centred cubic arrangement and the sodium chloride structure can be described as 'two interpenetrating face-centred cubic arrangements'.

3 is also incorrect as in the caesium chloride structure the Cs^+ and Cl^- ions are in 'two interpenetrating simple cubic arrangements'.

2 and **4** are correct and the key is **C**.

There are two interesting points hidden in the wording of **2**. These can be clearly seen if you have an electron density map available for sodium chloride or another ionic solid. Try to find such a map and note the following:

(1)   Ions are not spherical and do not have just one definite radius. Ions squash and expand to fill the space available in the lattice. They are at their smallest in the direction of the ions of opposite charge and at their largest in the direction of the ions of the same charge. This makes it necessary to select one direction in which to measure the radius of the ion. The chosen direction is towards the ion of opposite charge.

(2)   The electron density drops to very low values in the 'no-man's land' between the ions. It is not possible to locate where one ion stops and the other begins. X-ray diffraction studies will locate the nuclei and give an accurate value for the sum of the ionic radii. It is difficult to go beyond this and make a certain, precise division of the sum between the positive and negative ions.

11. Atoms of the elements of Group 3 have three electrons in their 'outer shell'. These atoms frequently bond to the atoms of other elements by sharing of electron pairs.

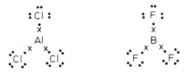

These molecules are planar and there is room for the formation of a bond with an extra atom. Both electrons of the new bond have to be supplied by a 'lone pair' on the extra atom. In the examples given below, $AlCl_3$ is bonding to itself using lone pairs on chlorine and BF_3 is shown bonding to the electron pair donor, ammonia.

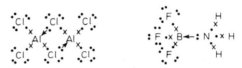

What is the shape of these molecules around the atoms of Al, B and N?

Only $AlCl_3$ and BF_3 of the molecules in the question can act as electron pair acceptors. The key is **C**. In NH_3 and CH_4 the central atom is already surrounded by four electron pairs and this is found to be the maximum for atoms of elements in the first short period.

There is some evidence that carbon can act as an electron pair acceptor. This occurs in the mechanism proposed for many organic reactions. One stage in these reactions involves a positive carbon ion, or carbonium ion, in which carbon is forming only three bonds,

$$\overset{\displaystyle |}{\underset{\displaystyle |}{-\,C+}}$$

This reacts with an electron pair donor such as hydroxide ions (*see* the answer to question 12 of test 15).

Carbon and nitrogen do not act as electron pair donors in their normal compounds. Atoms of higher atomic number in the same Group can, however, be surrounded by more than four electron pairs. Compounds of silicon and phosphorus, for example, can act as electron pair acceptors. In the examples given below, the central atom in the product is surrounded by six electron pairs.

$$SiF_4 \; + \; 2F^- \; \longrightarrow \; \left[\begin{array}{c} F \\ F \cdots \underset{F}{\overset{|}{Si}} \cdots F \\ | \\ F \end{array} \right]^{2-}$$

$$PCl_5 \; + \; Cl^- \; \longrightarrow \; \left[\begin{array}{c} Cl \\ Cl \cdots \underset{Cl}{\overset{|}{P}} \cdots Cl \\ | \\ Cl \end{array} \right]^-$$

What will be the shapes of the ions formed?

12. The bonding in each of the species is shown below

or   or   or

In **2** and **3** it is not possible to represent the species adequately by only one bonding structure. The electrons are not fixed in one arrangement. In the nitrate ion, for example, all bond lengths are identical and equivalent to 4/3 bonds each, all bond angles are exactly 120 degrees and the negative charge is distributed equally over the whole ion. The electrons in **2** and **3** are said to be delocalized. The key is **E**.

Two bonding structures are also shown for the ammonium ion, NH_4^+. How are the electrons arranged in these two forms of the ion? Is there any difference between the two arrangements of electrons? Whether your answer is "yes" or "no", does it influence the key in question 12?

13. Atoms of elements of higher atomic number than those in the first short period can form compounds where the atom has a share in more than four electron pairs. These atoms cannot be said to have acquired the electronic structure of a noble gas. SiF_6^{2-} and PCl_6^- have been given as examples at the end of question 11. SF_6 is another example. Statement **4** is incorrect.

The SF_6 molecule is octahedral (**2** incorrect) in which six equivalent bonds are formed be-

tween S and F (**1** correct). The oxidation number of sulphur is +6 and it is using six electrons in bonding so **3** is correct for SF_6. **1** and **3** are correct and the key is **B**.

Care is needed over the interpretation of the third statement in this question. It is true for SF_6, which is as far as the question tests, and for most other sulphur compounds but the statement is not true for all substances.

In sulphur itself, S_8, the element is using two electrons per atom in bonding but the oxidation number of sulphur is zero. This is because sulphur is bonding to no atom more electropositive or more electronegative than itself.

In the tetrathionate ion, $S_2O_3^{2-}$, one sulphur atom is using six electrons in bonding, the other only two (or zero if joined by a dative link to the central sulphur atom).

or

The oxidation number of sulphur is frequently quoted as +2 in this compound. This is an average oxidation number. The central sulphur atom is oxidation number +4 as it is using four electrons to bond to atoms more electronegative than itself. The oxidation number of the other sulphur atom is 0 as it is using no electrons to bond to atoms more electropositive or more electronegative than itself.

14. It is necessary to be certain about the bonds present in these molecules

Decane contains	C–H	C–C	
Trichloromethane contains		C–H	C–Cl
Tetrachloromethane contains			C–Cl

The absorption frequency of $3\cdot5 \times 10^{-6}$ m is present only in the two molecules that contain C–H as their common bond. It is a reasonable deduction that **1** is correct.

The absorption frequency of 7×10^{-6} m is present only in the one compound whose molecules do have C–C bonds. It is a reasonable deduction that **2** is correct.

The absorption frequency of $13 \cdot 5 \times 10^{-6}$ m is present only in the two molecules that contain C–Cl as their common bond. It is a reasonable deduction that **3** is correct.

The absorption frequency of $8 \cdot 3 \times 10^{-6}$ m is not present in both the compounds whose molecules contain C–H bonds. It is not a reasonable deduction that **4** is correct.

1, 2 and **3** are reasonable deductions so the correct key is **A**.

A problem remains with deduction **4**, however. If the $8 \cdot 3 \times 10^{-6}$ m frequency in $CHCl_3$ is not due to C–H, to what is it due? That it cannot be due to C–H is confirmed by its considerable strength of absorption. If $CHCl_3$ with only one C–H bond absorbs as strongly as this, the equivalent absorption with $C_{10}H_{22}$, which contains 22 C–H bonds, should be enormous. No such absorption occurs with decane anywhere near the frequency in question. The $8 \cdot 3 \times 10^{-6}$ m frequency in $CHCl_3$ cannot be due to C–Cl either as CCl_4 with more of these bonds shows no trace of absorption in the region of this frequency. There appear to be two possible explanations for the existence of this frequency. It may be due to a flexing of the molecule rather than a vibration along one of the lines joining nuclei, or it may be due to vibration in a new bond formed by intermolecular attraction between trichloromethane molecules.

15. The diffraction of X-rays by crystals is caused by the electrons present in the structure being investigated. The effect is so highly dependent on electron density that hydrogen atoms, with a lower electron density than any other atom, are rather difficult to spot and certainly very difficult to locate with precision. The best choice is **B** although this is rather too strongly worded.

    If hydrogen atoms are to be located as accurately as possible by a technique of this type, a beam of neutrons would be used instead of X-rays. Neutrons are unaffected by the electrons but are scattered by the nuclei in the crystal structure being investigated. Hydrogen nuclei interact strongly with neutrons.

16. Calculate the number of each type of particle present in the unit cell.

Black	1	[at the centre]
Small white	1	[1/8 at each corner, 8 × 1/8 in all]
Large white	3	[1/4 along each edge, 12 × 1/4 in all]

    The formula is

    $(Black)_1 (Small \ white)_1 (Large \ white)_3$.

    The only formula given which fits this is $EuTiO_3$, key **A**.

17. The difference between the KCl and KBr distances is 0·154 Å and is due to the greater radius of the Br^- ion. Assuming that each ion remains the same radius in different compounds, the RbBr distance should be 0·154 Å more than the RbCl distance. This suggests an answer of 3·439 Å which is, happily, one of those given, key **D**.

18. The shapes of these molecules are given in the answer to questions 1–4. The angle is at its greatest in **A** (180 degrees) and then in **B** (120 degrees). **C**, **D** and **E** resemble each other in that there are four electron pairs around the central atom. The angles in these three should be similar and close to the tetrahedral angle (109 degrees).

    It is found from experience of many molecules that lone pairs of electrons exert a slightly greater repulsional effect than do bonding pairs of electrons. The two lone pairs in H_2O make the H–O–H angle the smallest angle between covalent bonds, key **E**. The molecules happen to be arranged in the question in order of decreasing bond angle. Check in a data book the values for the bond angles in **C**, **D** and **E** to see what difference lone pairs make. You might also like to compare the bond angles in H_2O and H_2S and in NH_3 and PH_3 to see the effect of moving down a Group.

19. **A** cannot be correct as the very high values of the two quantities show that T must have a giant structure (metallic or giant covalent).

    B is correct as S must have a giant structure for the same reason that T has this type of structure. This should be the correct key but it is advisable to check the others just in case a point that the question is testing has been missed.

    C cannot be correct. A metal should have a high value for the molar enthalpy of vaporization and a boiling point well in excess of 373 K. What is the lowest value that the boiling point has for a metal?

D and **E** cannot be correct because the structures must go: small molecule–small molecule–small molecule–small molecule–giant structure –giant structure–giant structure–giant structure. It appears from the sequence that one period ends at P and the next begins at Q. The final four structures in more detail are likely to be: metallic–metallic–metallic (or giant covalent)–giant covalent.

20. The bond with the most ionic character will be the one with the greatest difference between the electronegativities of its component atoms. Of those between which choice has to be made, the best answer is **A**. See if you can find in a data book the percentage ionic character of a bond with electronegativity difference 1.0.

21. The lattice energy is concerned with the energy difference between one mole of solid lattice and the molar quantities of separate gaseous ions from which the lattice is built. This energy difference is shown in key **D**.

22. When gaseous methane is converted to separate gaseous atoms, four C—H bonds have to be broken.

$$CH_4(g) \rightarrow C(g) + 4H(g); \Delta H = +1660 \text{ kJ mol}^{-1}$$

The average C—H bond energy in methane, called in this question 'the bond energy term', is one-quarter of the enthalpy change in the above reaction. The value of the average C—H bond energy is ¼ × 1660 kJ mol^{-1}. The equation representing this change is shown correctly in key **A**.

23. The C – C single bond contains only one electron pair holding the carbon nuclei together compared with two electron pairs in the C = C double bond. This makes the C – C bond longer and weaker than the C = C bond. It is necessary to find in the table a bond length greater than 0.134 nm and a bond energy less than 610 kJ mol^{-1}. A pair of suitable values is given in key **D**.

24. The analytical technique that makes use of vibration within molecules is infra-red absorption spectroscopy, key **C**. Some spectra of this type are given in question 14.

25. Many compounds of transition elements contain unpaired electrons but it is extremely unusual to find compounds of non-transition elements with an unpaired electron. Nitrogen monoxide, key **A**, is one of these exceptional molecules.

The quick way to decide whether a molecule contains an unpaired electron is to look at whether the Group numbers of the elements that compose the molecule are odd or even. Groups 1, 3, 5, and 7 are odd. (Their total number of electrons is odd). Groups 2, 4, 6 and 8/0 are even. Find the total of odd and even atoms in the structure. If this total is odd, the structure contains an odd number of electrons. It is said to be an 'odd-electron molecule' or to contain an 'unpaired electron'. Some examples are given below.

		Sum
NO	1 × odd + 1 × even	odd
HNO$_3$	1 × odd + 1 × odd + 3 × even	even
NH$_3$	1 × odd + 3 × odd	even
N$_2$	2 × odd	even
HCN	1 × odd + 1 × even + 1 × odd	even

Use this method to check quickly whether the following molecules are odd-electron molecules : NO$_2$, N$_2$O$_4$, C1$_2$O and C1O$_2$. You should find that two of these are odd-electron molecules. It is usual for such molecules to be unstable, turning readily to molecules with an even number of electrons. Is this the case with the two molecules you have identified in the list of four?

26. In each of the structures given, the central atom (N, C or B) is surrounded by only three other atoms. Repulsion between electron pairs in the three bonds should make all these structures planar. This is not the case, however, in ammonia (key **C**). The lone pair on nitrogen provides a fourth electron pair. This repels the three electron pairs of the N–H bonds and forces them to take up a non-planar arrangement. The shapes of four of the molecules in this question is discussed in the chart on page 73.

27. Species that have the same total number of electrons as a chlorine atom are said to be 'isoelectronic' with chlorine. Cl^+ has one less electron, HCl has one more while S^- has the same number. The correct key is **E**.

   Ne$^-$ and OH$^-$ involve atoms far removed from chlorine in atomic number and so have a completely different total number of electrons. Is either of these two ions iso-electronic with a fluorine atom rather than with a chlorine atom?

28. The Cl^- ion has by far the largest ionic radius as it contains one complete shell of electrons more than the other four species. The other four are 'isoelectronic'. They each have the electronic configuration $1s^2, 2s^2, 2p^6$. The one with the smallest radius will be the one with the greatest attraction between the nucleus and the ten electrons that all possess. The greatest attraction will occur in the one with the largest nuclear charge, Mg^{2+}. The correct key is **A**.

29–31.
   The unit cell contains Sr^{2+} ions in the face-centred cubic arrangement (key **C** in question 29).

The structure must contain twice as many Cl^- ions as Sr^{2+} ions. This means that there must be twice as many Cl^- ions around each Sr^{2+} ion as there are Sr^{2+} ions around each Cl^- ion. The co-ordination numbers of Sr^{2+} and Cl^- must be in the ratio 2 to 1. This is only the case in **B** of question 30 which must be the correct key here.

The composition of the unit cell of $SrCl_2$ is:

8 Cl^- ions [all completely inside unit cell]

4 Sr^{2+} ions [$8 \times \frac{1}{8}$ at each corner +

$6 \times \frac{1}{2}$ at centre of each face]

The unit cell contains a total of 12 ions in a volume v cm^3. The number of ions in a volume V cm^3 will be $12 \times (V/v)$. But the volume V cm^3 also contains $3L$ ions as one mole of $SrCl_2$ contains L Sr^{2+} ions and $2L$ Cl^- ions.

$$3L = 12 \times \frac{V}{v}$$

$$\text{so } L = \frac{12}{3} \times \frac{V}{v}$$

$$= 4 \times \frac{V}{v}$$

which is key **D**.

Test 7

Intermolecular forces and solvation

Questions 1—5

The following list classifies a variety of substances according to composition:

A simple molecules only

B ions and simple molecules

C molecules (some of which are hydrogen-bonded)

D molecules (some of which are hydrogen-bonded and ions)

E ions and giant molecules

Select the category into which you would place a

1. solution of ammonia in water

2. mixture of sodium chloride and sand

3. solution of benzene in toluene

4. solution of acetic acid in benzene

5. suspension of potassium bromide in toluene

Questions 6—9 concern the following ways in which a system of two liquids may be classified:

A two immiscible or partially immiscible liquids

B a mixture that obeys Raoult's law

C a mixture that shows negative deviation from Raoult's law

D a mixture that shows positive deviation from Raoult's law

E two liquids that react together to form a stable compound

Select from A to E, the heading under which you would classify each system below.

6. Cyclohexane and cyclohexene

7. Ethanol and ethanoyl chloride (acetyl chloride)

8. Phenylamine (aniline) and water

9. Propan-l-ol and water

Directions summarized for questions 10 to 14

A	B	C	D	E
1,2,3 only correct	**1,3** only correct	**2,4** only correct	**4** only correct	Some other response or combination of responses is correct

10.

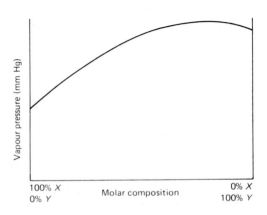

The above graph shows the variation in vapour pressure of a solution of two liquids X and Y as a function of molar composition.

It can be deduced that

1 pairs of liquids which give rise to this shape of curve evolve heat on mixing

2 the curve shows a positive deviation from Raoult's law

3 Y has a higher boiling point than X

4 the two liquids could be ethanol and cyclohexane

11. Mixtures of acetone and trichloromethane show a variation in boiling point similar to the following:

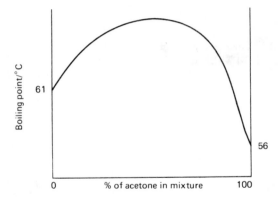

Which of the following statements is (are) correct?

1 Mixtures of acetone and trichloromethane show positive deviation from Raoult's law.

2 The forces of attraction acting between molecules of acetone and trichloromethane in a mixture of these liquids are greater than those acting between the molecules in pure acetone.

3 Pure acetone can be obtained by the careful fractional distillation of any mixture of acetone and trichloromethane.

4 When acetone and trichloromethane are mixed, the enthalpy change is negative.

12. When 4·24 g of anhydrous lithium chloride (LiCl = 42·4) is dissolved in water to make 500 cm³ of solution,

1 the solution formed is 0·05M with respect to Cl⁻

2 the lithium ions would be hydrated, the water molecules surrounding the lithium ions in the following manner

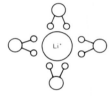

3 the solution formed would have a pH of about 3·5

4 some hydrogen bonds in the water would be broken

13. The energy of sublimation of solid helium is 0·105 kJ mol^{-1} whereas that of ice is 46·9 kJ mol^{-1}. Which of the following statements help to explain these facts?

    1 Only van der Waals forces are present between helium atoms.

    2 There is extensive hydrogen bonding in liquid water.

    3 Helium atoms are very small with few electrons.

    4 The covalent bonds in water molecules are very strong.

14. If two liquids are mixed and form an ideal solution

    1 there is no enthalpy change

    2 the vapour pressure of the mixture is the sum of the partial vapour pressure of the components

    3 the total volume of the mixture is equal to the sum of the volumes of the two components

    4 the molecules of neither component contain hydrogen atoms

---

Directions for questions 15 to 21. Each of the questions or incomplete statements in this section is followed by five suggested answers. Select the best answer in each case.

---

15. If a certain metal chloride is added to cold water, the solution becomes very hot and may even boil. This solution is found to conduct electricity.
    Which of the following is most likely to be responsible for the evolution of the heat?

    A Dissociation

    B Sublimation

    C Hydration

    D Ionization

    E Ionization and dissociation together

16. The energy evolved when one mole of gaseous calcium ions is hydrated according to the equation

    $$Ca^{2+}(g) + aq \rightarrow Ca^{2+}(aq)$$

    is greater than the corresponding value for barium (Ba^{2+}) ions because the

    A ionization energy of calcium is greater than that of barium

    B atomic radius of calcium is greater than that of barium

    C lattice energy of calcium oxide is greater than that of barium oxide

    D ionic radius of Ca^{2+} is less than that of Ba^{2+}

    E solubility of calcium hydroxide in water is less than that of barium hydroxide

17. Which of the following compounds would you expect to have the largest dipole moment in the vapour phase?

    A CH_4

    B NH_3

    C H_2O

    D NO

    E HF

18. At 20 °C the vapour pressure of water is 18 mmHg. Assuming that the system obeys Raoult's law, what is the partial vapour pressure (in mmHg) of water in a solution containing 72 g of water and 32 g of methanol at 20 °C? (H = 1, C = 12, O = 16)

    A 18

    B 18 × 4/5

    C 18 × 5/4

    D 18 × 4

    E 18 × 5

19. Consider the following molecular interactions in the liquids:

$$H_2O \ldots H_2O \qquad HF \ldots HF$$
$$\text{I} \qquad\qquad\qquad \text{II}$$

$$CH_3COCH_3 \ldots CH_3COCH_3$$
$$\text{III}$$

$$C_2H_5OH \ldots C_2H_5OH$$
$$\text{IV}$$

If the above pairs of molecules are arranged in the order of the strengths of the interactions between them, putting the pair with the *strongest interaction first*, the order would be

A I, II, III, IV

B II, III, I, IV

C II, I, IV, III

D II, I, III, IV

E I, III, IV, II

20. The following substances have approximately the same relative molecular mass. Which is likely to have the highest boiling point?

A $CH_3CH_2CH_2CH_2CH_3$

B $C_2H_5OC_2H_5$

C $CH_3CH_2CH_2SH$

D $(CH_3)_2NC_2H_5$

E $CH_3CH_2CH_2CH_2OH$

21. The total number of solvent molecules loosely or firmly bound to a given ion is known as the *solvation number*. Consider the following data:

Ion	Approximate solvation number
X	5
Y	15
Z	25

X, Y and *Z* are most likely to be ions of

A Group 1 metals arranged respectively in order of increasing atomic number

B Group 2 metals arranged respectively in order of increasing atomic number

C Group 7 elements arranged respectively in order of increasing atomic number

D Group 7 elements arranged respectively in order of decreasing atomic number

E the first three elements respectively in a short period of the Periodic Table

22. Sodium chloride is likely to be most soluble in

	Dipole moment $/10^{-3}$ Cm	Relative permittivity
A Tetrachloromethane	0	2
B Ammonia	3	22
C Methanol	6	33
D Hydrogen fluoride	7	84
E Propanone	10	21

23. The enthalpy of hydration of sodium chloride can be represented by the equation

A $Na(s) + \frac{1}{2}Cl_2(g) + aq \rightarrow$
$$Na^+(aq) + Cl^-(aq)$$

B $Na^+(g) + Cl^-(g) + aq \rightarrow$
$$Na^+(aq) + Cl^-(aq)$$

C $Na(s) + HCl(aq) \rightarrow$
$$Na^+(aq) + Cl^-(aq) + \frac{1}{2}H_2(g)$$

D $NaCl(s) + aq \rightarrow Na^+(aq) \quad Cl^-(aq)$

E $NaCl(g) + aq \rightarrow Na^+(aq) + Cl^-(aq)$

24.

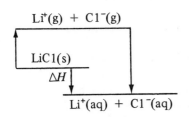

The energy change ΔH in the energy level diagram above may be described as the

A heat of solution

B heat of formation

C heat of hydration

D heat of ionization

E lattice energy

25. For a salt to have a positive heat of solution in water

    A its lattice energy must be more negative than its hydration energy

    B its hydration energy must be more negative than its lattice energy

    C its standard heat of formation must be more negative than its hydration energy

    D its standard heat of formation must be more negative than its lattice energy

    E its hydration energy must be more negative than its standard heat of formation

26.

$$CH_3\,CH_2\,CH_2\,CH_3 \qquad CH_3 - \overset{\displaystyle CH_3}{\underset{\displaystyle H}{\overset{|}{\underset{|}{C}}}} - CH_3$$

       W   *X*

$$CH_3\,CH_2\,CH_2\,CH_2\,Br \qquad CH_3 - \overset{\displaystyle CH_3}{\underset{\displaystyle CH_3}{\overset{|}{\underset{|}{C}}}} - Br$$

          Y   *Z*

The boiling points of the compounds *W*, *X*, *Y* and *Z* increase in the order

A *WXYZ*

B *YZWX*

C *XWZY*

D *XZWY*

E *ZXYW*

27. The following substances have very similar molar masses. The one with the lowest boiling point is most likely to be

    A $CH_3 - \underset{\displaystyle O}{\overset{\displaystyle \|}{C}} - OH$ **D** $CH_3 - \underset{\displaystyle CH_3}{\overset{|}{C}H} - CH_3$

    B $CH_3 - \underset{\displaystyle NH_2}{\overset{|}{C}H} - CH_3$ **E** $CH_3 - \underset{\displaystyle O}{\overset{\displaystyle \|}{C}} - CH_3$

    C $CH_3 - \underset{\displaystyle OH}{\overset{|}{C}H} - CH_3$

Questions 28–30

The following experiment is intended to investigate the change in volume on mixing acetic acid and benzene.

Directions

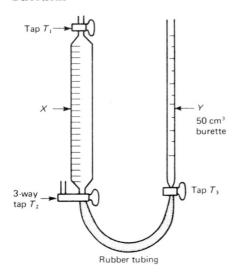

Tap T_1 →

X →

Y ←
50 cm³
burette

3-way tap T_2

Tap T_3

Rubber tubing

Set up the apparatus as shown in the diagram with the graduated tube X about half-filled with acetic acid, tap T_1 closed and tap T_2 opened to connect the rubber tubing with the atmosphere.

Pour benzene into Y, with tap T_3 open, until liquid emerges from tap T_2. Close taps T_3 and T_2, fill Y to the zero mark with benzene.

Raise Y, open all the taps carefully to join the rubber tubing and X and allow a suitable volume of benzene to enter X.

Close all the taps. Shake X to mix the liquids thoroughly and open tap T_1. Measure the volume of liquid in X and take a volume reading in Y.

Repeat, adding further volumes of benzene from Y to X and take similar readings.

From the results it is intended to draw a graph of the quantity (Volume of mixed liquids −Total volume of separate liquids in mixture) against the percentage composition by volume of the mixture.

28. In order to be able to plot this graph, it would be necessary to measure

   A the volume of the "dead space" in Y

   B the volume of the rubber tubing

   C both the volume of the dead space in Y and the volume of the rubber tubing

   D the total volume of acetic acid in X

   E the total volume of benzene in Y

29. The apparatus could be simplified without loss of accuracy by using

   A an open tube instead of tap T_1

   B an open tube instead of tap T_3

   C an ordinary two-way tap instead of tap T_2

   D open tubes in place of both T_3 and T_1

   E an ordinary burette in place of X

30. Denoting the quantity (Volume of mixed liquids−Total volume of separate liquids in mixture) by ΔV, the graph would be of the form

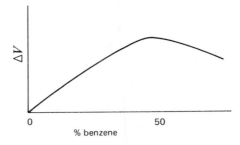

If acetic acid molecules are represented by P and benzene molecules by Q, then a possible explanation for these results is that the

   A $P–Q$ bond is weaker than either the $Q–Q$ bond or the $P–P$ bond

   B $Q–Q$ bond is weaker than the $P–P$ bond

   C $P–P$ bond is weaker than the $Q–Q$ bond

   D $P–Q$ bond is stronger than either the $Q–Q$ bond or the $P–P$ bond

   E $P–Q$ bond is stronger than the $P–P$ bond

Test 7 Answers

1. Ammonia and water molecules are hydrogen-bonded to themselves and to each other. Acid–base reactions in the solution produce ions.

$$H_2O(l) + H_2O(l) \rightleftharpoons H_3O^+(aq) + OH^-(aq)$$

$$H_2O(l) + NH_3(aq) \rightleftharpoons NH_4^+(aq) + OH^-(aq)$$

The correct key is **D**. 'Some' should really be replaced by 'all' for this mixture.

2. Sodium chloride has a giant structure of ions, and sand, silicon dioxide, has a giant covalent structure. The best answer is **E**.

3. Benzene and toluene are both composed of small covalent molecules with similar shapes and bonding. They mix in all proportions to form a near-ideal mixture in which the weak forces between different molecules are comparable in strength with the weak forces between the component molecules. No hydrogen-bonding is possible so the key is **A**.

4. Benzene is composed of non-polar molecules and is not an ionizing solvent for substances dissolved in it.
    Acetic acid (ethanoic acid) is a polar molecule with extensive hydrogen-bonding between molecules. This shows in the high boiling point for a substance of relative molar mass 60. Butane, C_4H_{10}, with RMM 58 is a gas at room temperature. The RMM of acetic acid vapour at the boiling point is close to 120, showing that hydrogen-bonding continues in the vapour state.
    When dissolved in water, acetic acid hydrogen-bonds to water molecules and also is weakly ionized. Neither of these reactions is possible when acetic acid dissolves in benzene. The fact that acetic acid dissolves shows that its molecules must interact with benzene molecules. This interaction will involve dipole–induced dipole and van der Waals forces (*see* the answer to question 13). As it cannot hydrogen-bond to the solvent, acetic acid remains hydrogen-bonded to itself in benzene, mainly as dimers. The correct key is **C**.

5. Potassium bromide has a giant structure of ions, toluene is composed of simple molecules. There is no interaction between them and the key is **B**.

6. Cyclohexane and cyclohexene are very similar. They are covalently bonded with very little polarity. Each is composed of non-planar rings of six carbon atoms. The formulae and relative molar masses are almost the same. The two molecules are about as similar as two molecules could be without being identical. Such a mixture would be classified as obeying Raoult's law, key **B**.
    A mixture would only obey Raoult's law perfectly if the molecules in it could not recognize whether they were surrounded by their own kind or by molecules of the other component. Such a pair of molecules would have to be as near identical as possible without being the same. Two possible systems where this is the case are $D_2^{16}O/H_2^{18}O$ and CH_3OD/CH_2DOH. [What would the vapour pressure–mole fraction curve look like for such a mixture?] Even these mixtures would be expected to show a very small positive deviation from Raoult's law as the different molecules would not pack as perfectly as molecules of only one kind.
    On moving to pairs that are less identical, such as two isomers of C_5H_{12}, each type of molecule will disrupt the packing of the other in a significant way and positive deviations from Raoult's law will become more noticeable. A positive deviation is expected for all mixtures unless a new, strong interaction is present. This new interaction can appear as hydrogen-bonding (trichloromethane–propanone) or as an acid–base reaction (pyridine–ethanoic acid).

7. There is such a strong interaction between these two substances that they form new compounds in a vigorous, often explosive, reaction. The key is **E**.

85

8. These molecules are so dissimilar in size and shape that it is asking too much for each to give up its own, special hydrogen-bonded packing to make room for the other. The liquids are almost immiscible, key **A**.

9. Water and alcohols have rather similar structures and, if the carbon chain is not too long, the liquids mix in all proportions. This is the case with propan-1-ol and water. The hydrogen-bonded packing of each component is disrupted by mixing so that the vapour pressure of each is increased despite the formation of new hydrogen-bonds between propan-1-ol and water. The mixture shows a positive deviation from Raoult's law, key **D**.

10. The curve shows a positive deviation from Raoult's law; 2 is a correct deduction. The two liquids could be ethanol and cyclohexane where the dissimilar liquids disrupt each other's packing and intermolecular forces without introducing any new, stronger intermolecular forces on mixing; 4 is a correct deduction.

   At the temperature of the graph, Y has a higher vapour pressure than X so it can be deduced that Y will have a lower boiling point than X; 3 is an incorrect deduction.

   There is a problem over the first deduction. The positive deviation from Raoult's law shows that stronger intermolecular forces are not present after mixing. Heat will certainly not be evolved for bonding reasons. You are expected to stop here and recognize that 1 is meant to be an incorrect deduction so that the key is **C**. There is another way, however, in which heat can be evolved on mixing. This occurs if, on mixing, there is a decrease in the number of ways in which the molecules can hold internal energy of movement, vibration and rotation. It is not possible to decide from the graph whether this will happen so 1 is still an incorrect deduction and the key is still **C**.

11. The boiling point—mole fraction curve shows a strong maximum so the vapour pressure—mole fraction curve is certain to have a minimum. This is a negative deviation from Raoult's law. The first statement is incorrect. This deduction can only really be made with certainty for vapour vapour pressures in the region of 60 °C, a temperature range covered by the boiling point curve. To understand why this is so, you would need to ask your teacher to help you construct a three-dimensional boiling point—vapour pressure—mole fraction diagram, possibly using Perspex sheets or plaster of Paris.

   The minimum in the vapour pressure curve indicates stronger intermolecular forces between the components than within each component. 2 is correct.

   Pure acetone can only be obtained by fractional distillation of a mixture of acetone and trichloromethane which already contains over about 50 per cent acetone. 3 is incorrect. If the mixture contains under about 50 per cent acetone, pure trichloromethane can be obtained by fractional distillation. The mixture remaining in the flask after both distillations would be the mixture of maximum boiling point containing about 50 per cent acetone.

   The intermolecular forces disturbed on mixing are weaker than the new intermolecular forces. The difference in the strength of the intermolecular forces contributes to a negative enthalpy change. It is not possible to *predict* from this whether the final enthalpy change will be negative because no information is given about the change in the number of ways in which the molecules can hold their internal energy of movement, vibration and rotation (*see* also the answer to question 10). If you were meeting this problem as a new situation, you would have to rely on the information given by the boiling point curve which indicates stronger bonding on mixing and hence a negative enthalpy change. You would have to regard the fourth statement as correct and experiment would show that this is, in fact, the case.

   2 and 4 are correct, key **C**.

12. The solution of LiCl contains 0·1 mole of compound in 500 cm^3 which makes the solution 0·2 M. Because of complete ionization, the molarity of Cl$^-$ will also be 0·2. 1 is incorrect.

   Water molecules are likely to be arranged in an octahedral rather than planar structure. The negative ends of the water molecules (oxygen) should be closest to the positive ion. 2 is correct.

   Chlorides of Group 1 elements are not hydrolysed in water. The pH of such solutions would be close to 7. 3 is incorrect.

   It is certain that some hydrogen-bonds in

water would be broken so that hydration of lithium and chloride ions could take place. It is this hydration which enables an ionic compound to dissolve in water. 4 is correct and the key is **D**.

13. van der Waals forces make a significant contribution to the intermolecular attraction between all molecules even if the forces of dipole–dipole, dipole–induced dipole and hydrogen-bonding are also present. In a noble gas, such as helium, the only possible attraction of any magnitude is van der Waals and this helps to explain the low sublimation energy of helium compared with ice.

    The van der Waals forces depend on the size and number of electrons in the interacting particles. The small size and small number of electrons contribute to the low sublimation energy of helium. Only the first and third statements help explain the information given in the question and the key is **B**.

    It is the hydrogen-bonding in ice rather than in liquid water which helps explain the high value for the sublimation energy of ice. Statement 2 does not help. Sublimation in ice does not involve breaking the strong covalent bonds within the water molecule so statement 4 does not help either.

14. If a mixture really is ideal, there will be no change in the intermolecular forces and no change in the way molecules hold their internal energy. 1 is correct. 2 is a statement of Raoult's law and will be correct for an ideal mixture. Molecules in an ideal mixture fail to recognize that they are in contact with molecules that are not all their own kind. One of the consequences of this is the additive nature of the volume of each component. 3 is correct, as are 1 and 2, and the key is **A**.

    4 is incorrect. It does not matter if the molecules of the components contain hydrogen atoms so long as these are unable to hydrogen-bond. If it had continued ' . . . . . joined to atoms of nitrogen, oxygen or fluorine' this would have excluded the possibility of hydrogen-bonding but would not have made certain that the mixture was ideal. The absence of hydrogen would not prevent the mixture's deviating from ideality because of differences in size, shape and polarity of its components.

15. Some metal chlorides ($AlCl_3$ and $SnCl_4$, for example) are hydrolysed by water in a strongly exothermic reaction which produces a conducting solution of hydrogen ions and chloride ions. Hydrolysis, however, is not among the changes in the question. Hydrolysis of metal chlorides involves hydration. Hydration is also responsible for the strongly exothermic reactions when anhydrous magnesium chloride and calcium chloride dissolve in water to produce a conducting solution of metal ions and chloride ions. The best choice is **C**.

16. The hydration energy of $Ca^{2+}(aq)$ is greater than the value for $Ba^{2+}(aq)$ for the reasons stated in **D**. The negative ends of the water molecules, where the lone pairs on oxygen are located, can approach the smaller calcium ion more closely and produce stronger bonding and greater evolution of energy.

    The hydration of gaseous ions cannot be related to the statements made in **A**, **B** and **C**. An explanation of the statement made in **E** could include the relative hydration energies of gaseous Ca^{2+} and Ba^{2+}. Note, however, that the relative hydration energies alone would suggest that the solubility of calcium hydroxide should be greater than that of barium hydroxide. Even if the solubilities were the other way round, the hydration energies are part of the explanation for **E** rather than **E** being an explanation for the hydration energies.

17. CH_4 has no dipole moment as the molecule is symmetrical. The choice is between the other four which are all polar. The high electronegativity of fluorine and its combination with electropositive hydrogen makes HF the molecule that would be expected to have the largest dipole moment, key **E**.

18. The mixture contains $\dfrac{72\ g}{18\ g\ mol^{-1}}$

            = 4 moles of water

    and $\dfrac{32\ g}{32\ g\ mol^{-1}}$

            = 1 mole of methanol

The mole fraction of water is $4/(4 + 1)$ which is $4/5$. If Raoult's law is obeyed, vapour pressure

= $18 \times (4/5)$ mmHg. The correct key is **B**. Do you think that this mixture really will obey Raoult's law?

19. I, II and IV will all form hydrogen-bonds but III will not, so III will come last. This makes the only possible choice key **C** on this observation alone. The other three are in the expected order of hydrogen-bonding strength in **C**.

You probably know that the F—H·······F hydrogen-bond is by far the strongest hydrogen-bond, approaching a normal covalent bond in strength. The O—H·····O hydrogen-bond is considerably weaker. This fits the order given in **C**.

Check the boiling points of HF and H_2O in a data book. Does the compound containing the stronger hydrogen-bonds boil at a higher temperature? If the boiling points are unexpected, they can be explained if you can see the significance of the following two pieces of information.

(1) Oxygen, with its two lone-pairs, can form two hydrogen-bonds per atom while fluorine, with only one lone-pair, can form only one hydrogen-bond per atom.

(2) The relative molar mass of hydrogen fluoride vapour at the boiling point is 70 while the relative molar mass of water vapour at the boiling point is 18.

20. Hydrogen-bonding is only possible in **E** which is likely to have the highest boiling point.

21. The interaction of solvent molecules with ions is favoured by small size and high charge on the ion. For these reasons, solvation number increases on going up Groups 1, 2 and 7 and across a short period from Group 1 to Group 3. It is this increase across a short period which correctly links key **E** with the given solvation numbers.

22. The dissolving of an ionic solid can be thought of as taking place in two stages. The first stage involves the separation of each ion from its neighbours of opposite charge. This separation is made much easier if the solvent has a high relative permittivity. In hydrogen fluoride, for example, the addition of this solvent make the attractive forces between the ions 84 times *lower*. The next-best solvent from this point of view is methanol where the attractive force between ions of opposite charge is 33 times lower when the solvent is present.

The second stage in the dissolving of an ionic solid is the solvation of each ion by polar solvent molecules. The positive ends of several solvent molecules cluster round each negative ion, while the negative ends of several solvent molecules cluster round each positive ion. The solvation of ions is favoured by high polarity in the solvent molecule, this polarity being indicated by the dipole moment in the table in the question. The most polar solvents here are methanol, hydrogen fluoride and propanone.

The best solvent for an ionic substance, such as sodium chloride, is most likely to be hydrogen fluoride which has by far the highest relative permittivity and the second highest dipole moment. The key is **D**.

Can you write structural formulae for each of the molecules in the question and show the positions of the atoms that have a small positive or negative charge, $\delta+$ or δ^-? Your structural formula for tetrachloromethane should show four atoms with a small negative charge and one atom with a small positive charge. Why does this molecule have zero dipole moment?

23. The energy released when one mole of a substance, in the form of separate gaseous ions, is solvated is called the 'solvation energy'. When the solvent is water, the term is 'hydration energy'. The hydration energy of sodium chloride will be concerned with the change from gaseous sodium and chloride ions to the aqueous ions and this is shown in equation B. The change is shown in the energy level diagram of question 24 where the energy change on the right of the diagram is the hydration energy of lithium chloride.

24. The energy change represented by ΔH in the diagram is the heat of solution, key **A**.

25. In the diagram in question 24, the hydration energy is larger than the lattice energy. This makes the energy level of the aqueous ions lower than that of the solid and gives rise to a negative heat of solution. If the lattice energy is larger than the hydration energy, the energy level of the aqueous ions is higher than that of the solid. This is the situation when the heat of solution is positive. The correct key is **A** rather than **B**.

26. Molecules Y and Z contain a bromine atom which gives them a larger total number of electrons and also makes them more polar than molecules W and X. Both these factors contribute to stronger intermolecular forces, giving Y and Z higher boiling points than W and X. The correct key must be **A** or **C**.

    The attractive forces between unbranched chain molecules is slightly greater than that between their branched chain isomers. This gives unbranched chain isomers slightly higher boiling points. Y will have a higher boiling point than Z and W will have a higher boiling point than X. The order of boiling points is XWZY, key **C**.

27. Molecules A, B and C contain a hydrogen atom attached to a small, highly electronegative nitrogen or oxygen atom. This means that hydrogen-bonding will make an important contribution to the intermolecular forces in these compounds. The presence of nitrogen or oxygen atoms in A, B, C and E will give these molecules large, permanent dipoles which will also contribute to strong intermolecular forces.

    Molecule D, however, is neither strongly polar nor capable of forming hydrogen-bonds. This compound will possess the weakest intermolecular forces and hence the lowest boiling point. The correct key is **D**.

28. **A**, **B**, **C** and **E** are unnecessary as the volume of benzene run into the acetic acid is given by the difference between the burette readings.

The volume of acetic acid in X is required and **D** is the correct key.

29. It is necessary to be able to shake the mixture of liquids in tube X so there must be a tap T_1. **A** and **E** are incorrect choices for simplification. It must be possible to fill the rubber tubing completely, removing any trapped air, so the three-way tap T_2 is essential (**B** an incorrect choice). Tap T_3 is unnecessary but tap T_1 is required for shaking purposes so the correct choice for simplification is **B** rather than **D**.

30. The graph shows that the mixture of acetic acid and benzene molecules occupies a greater volume than the sum of the volumes of the components. This suggests a weakening of the intermolecular forces on forming the mixture and the only explanation, among those given, which fits is key **A**.

    Answers in this test have already discussed how care has to be taken over drawing conclusions from temperature changes on mixing pairs of liquids. It is also necessary to be careful over the interpretation of volume changes on mixing. A good example of this occurs on mixing water and ethanol. The mixture shows a positive deviation from Raoult's law which shows that the intermolecular forces are weaker in the mixture than in the original components. As the mixture is made up, however, there is a volume decrease and a temperature rise. Both these observations appear to suggest stronger bonding in the mixture, in conflict with the vapour pressure data.

    The resolutions of this problem is connected with the structure of water. Water, like ice, has a very open, hydrogen-bonded structure with considerable 'empty' space. When this structure is partly broken up by the formation of new bonds with ethanol, some of this space is lost and the mixture contracts.

    Another feature of the water structure is the large number of ways in which it can hold vibrational energy. This shows in the high molar heat capacity of water. When the water structure is partly broken up by its molecules being forced to interact with ethanol, some of the ways of holding vibrational energy disappear. The

energy that was held in these vibrations is transferred to a general increase in the magnitude of
every type of energy held in the mixture. This
is another way of saying that the lost vibrations
appear as heat, thus accounting for the
temperature rise.

Test 8   Reaction rates

Questions 1–3 refer to five lettered graphs of a quantity Y plotted against a quantity X.

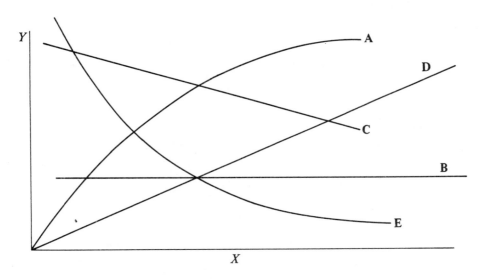

Choose the graph which best shows the relationship between Y and X in each of the following questions. Each graph may be used once, more than once, or not at all.

1. For the reaction

$$H_2(g) + I_2(g) \rightarrow 2HI(g)$$

Y is the concentration of HI and X is the time from the beginning of the reaction.

2. For the reaction

$$CH_3CO_2C_2H_5(aq) + OH^-(aq) \rightarrow CH_3CO_2^-(aq) + C_2H_5OH(aq)$$

Y is the concentration of ethyl acetate and X is the time after mixing equimolar proportions of the reactants.

3. Y is the rate of reaction, and X is the concentration of reactant, in a reaction which is zero order with respect to that reactant.

A	B	C	D	E
Directions summarized for questions 4 to 9				
1,2,3 only correct	1,3 only correct	2,4 only correct	4 only correct	Some other response or combination of responses is correct

4. Aqueous solutions of bromine and propanone (acetone) react according to the equation

$$CH_3COCH_3 (aq) + Br_2 (aq) \rightarrow$$
$$CH_3COCH_2Br(aq) + HBr(aq)$$

This reaction is zero order with respect to bromine. It may be inferred that the

1   reaction rate is constant

2   slowest reaction step involves bromine

3   bromine acts as a catalyst

4   rate determining step does NOT involve bromine

5. The bromination of an alkane takes place in the following stages:

I $Br_2 \qquad\qquad \rightarrow 2Br\cdot$ \qquad Initiation

II $RCH_3 + Br\cdot \rightarrow RCH_2\cdot + HBr$ $\Big\}$ Propaga-

III $RCH_2\cdot + Br_2 \rightarrow RCH_2Br + Br\cdot$ $\Big\}$ tion

Given the bond energy terms (in kJ mol^{-1})

C–H 409;   Br–Br 192;   H–Br 367;
C–Br 284

which of the following statements is/are correct?

1   The initiation stage is exothermic.

2   Heat is emitted in stage II.

3   Heat is absorbed in stage III.

4   The propagation (stages II and III) is exothermic.

6. The rate of the reaction between sodium hydroxide solution and ethyl propanoate

$$C_2H_5CO_2C_2H_5 (l) + OH^- (aq) \rightarrow$$
$$C_2H_5CO_2^-(aq) + C_2H_5OH(aq)$$

is first order with respect to both ethyl propanoate and hydroxide ions. It is true to say that

1   halving the concentration of either of the two reactants independently will halve the rate of reaction

2   the reaction is first order overall

3   the rate of reaction is proportional to the concentration of ethyl propanoate

4   halving the concentrations of both reactants simultaneously will halve the rate of reaction

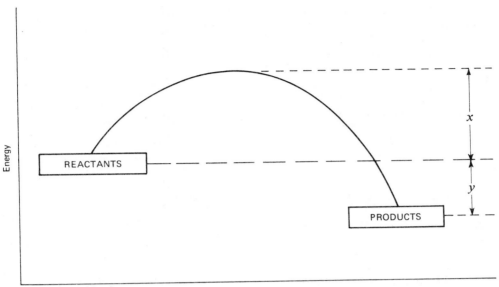

Reaction path

7. The energy profile shown above related to
the reaction:

$$CO(g) + NO_2(g) \rightleftharpoons CO_2(g) + NO(g)$$

Which of the following statements follow from
this?

1 The activation energy of the reverse reaction
is $(x + y)$.

2 ΔH for the reverse action is x.

3 The forward reaction is exothermic.

4 Both forward and reverse reactions are
second order overall.

8. An investigation into the initial rate of hydrolysis of urea by the enzyme urease produced
the graph shown in the next column. The
enzyme concentration was constant in all
experiments.

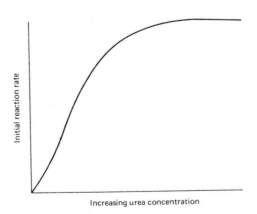

Increasing urea concentration

These results suggest that

1 at low urea concentration the initial rate
depends on urea concentration

2 at high urea concentrations the initial rate
is independent of the concentration of urea

3 at high urea concentrations the reaction
at the start is zero order with respect to
urea

4 the reaction rate increases as hydrolysis
proceeds

9. Hydrogen peroxide oxidizes hydriodic acid
according to

$$H_2O_2(aq) + 2H^+(aq) + 2I^-(aq) \rightarrow$$
$$I_2(aq) + 2H_2O(l).$$

The rate of the reaction could be investigated
by measuring the large changes in the

1   volume of the system

2   intensity of colour of the system

3   optical rotation of the system

4   electrical conductivity of the system

Directions for questions 10 to 15. Each of the
questions or incomplete statements in this section
is followed by five suggested answers. Select the
best answer in each case.

---

10. For the reaction $X + Y \rightarrow Z$ the rate expression
is

$$\text{Rate} = k[X]^2[Y]^{1/2}$$

If the concentrations of X and Y are both
increased by a factor of 4, by what factor will
the rate increase?

A   4

B   8

C   16

D   32

E   64

11. Which graph represents the plot of rate of
reaction against time for a first order reaction?

A   Rate
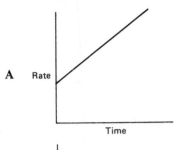

B   Rate
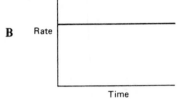

C   Rate
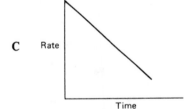

D   Rate
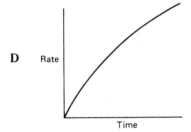

E   Rate
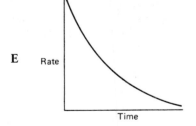

12. When the rate of the reaction

    X(aq) + Y(aq) → XY(aq)

    is investigated experimentally using equal concentrations of X and Y in the presence of sodium hydroxide, it is found that

    Rate = k[X(aq)] [OH⁻(aq)]

    Which of the following mechanisms for the reaction taking place under the conditions specified is consistent with this rate equation?

    A Y(aq) + OH⁻(aq) $\xrightarrow{\text{(slow)}}$ YOH⁻(aq)

    YOH⁻(aq) + X(aq) $\xrightarrow{\text{(fast)}}$

    XY(aq) + OH⁻(aq)

    B Y(aq) + OH⁻(aq) $\xrightarrow{\text{(fast)}}$ YOH⁻(aq)

    YOH⁻(aq) + X(aq) $\xrightarrow{\text{(fast)}}$ XYOH⁻(aq)

    XYOH⁻(aq) $\xrightarrow{\text{(slow)}}$

    XY(aq) + OH⁻(aq)

    C X(aq) + OH⁻(aq) $\xrightarrow{\text{(fast)}}$ XOH⁻(aq)

    XOH⁻(aq) + Y(aq) $\xrightarrow{\text{(fast)}}$ XYOH⁻(aq)

    XYOH⁻(aq) $\xrightarrow{\text{(slow)}}$

    XY(aq) + OH⁻(aq)

    D X(aq) + OH⁻(aq) $\xrightarrow{\text{(fast)}}$ XOH⁻(aq)

    XOH⁻(aq) + Y(aq) $\xrightarrow{\text{(slow)}}$

    XY(aq) + OH⁻(aq)

    E X(aq) + OH⁻(aq) $\xrightarrow{\text{(slow)}}$ XOH⁻(aq)

    XOH⁻(aq) + Y(aq) $\xrightarrow{\text{(fast)}}$

    XY(aq) + OH⁻(aq)

13. The rate of reaction between X and Y is third order overall. Which of the following rate equations must be INCORRECT?

    A Rate = $k[X][Y]^3$

    B Rate = $k[X][Y]^2$

    C Rate = $k[X]^2[Y]$

    D Rate = $k[X]^2[Y][Z]^0$

    E Rate = $k[X]^0[Y]^3$

14. When a dilute, aqueous solution of potassium permanganate is run from a burette into a flask containing dilute, aqueous oxalic acid and dilute sulphuric acid, the rate of the reaction suddenly increases considerably as more potassium permanganate is added. The reason for this is that

    A the manganese(II) ions produced catalyse the reaction

    B the pH of the solution in the flask increases

    C a definite minimum concentration of permanganate ions is necessary before the reaction will proceed

    D the reaction is exothermic and the heat energy liberated affects the rate

    E the sulphuric acid removes water and so causes the reaction to proceed more rapidly to completion

15. Two colourless substances X and Y react to give a coloured substance Z. The times (t) taken for various initial concentrations of X and Y to produce a certain colour intensity are recorded in the table.

$[X]$/mol dm^{-3}	$[Y]$/mol dm^{-3}	t/s
0.05	0.05	44
0.05	0.10	22
0.10	0.05	44

Which rate equation is consistent with these results?

A   Rate $= k[Y]^{1/2}$

B   Rate $= k[Y]$

C   Rate $= k[Y]^2$

D   Rate $= k[X][Y]$

E   Rate $= k[X][Y]^2$

Questions 16–20 concern an experiment to determine the initial rate of reaction between oxidizing agent ammonium peroxydisulphate $[(NH_4)_2 S_2 O_8]$ and potassium iodide.

A series of experimental runs was carried out. In these exactly 10 cm^3 of 5×10^{-3} M sodium thiosulphate together with exactly 3 drops of a starch solution were placed in a conical flask and 20 cm^3 each of the $(NH_4)_2 S_2 O_8$ solution and KI solution were poured into this together. The flask was swirled and a stop clock started. The time taken for the solution to darken was noted.

The initial concentrations of the $(NH_4)_2 S_2 O_8$ and KI solutions in the mixture together with the times to darken, for the various experimental runs, are given below.

Initial concentrations/mol dm^{-3}		Times to darken/s
$(NH_4)_2 S_2 O_8$	KI	
0.10	0.20	35
0.05	0.20	69
0.03	0.20	103
0.10	0.10	70
0.10	0.067	104

16. The darkening of the solution was due to the

A   formation of a complex ion from the peroxydisulphate

B   formation of an iodine–thiosulphate compound

C   formation of a polysaccharide–iodine complex

D   oxidation of sodium thiosulphate

E   precipitation of colloidal sulphur

17. The experiment was carried out by visual inspection. Which of the following methods could also be used?

A   Polarimetry

B   Colorimetry

C   Dilatometry

D   Titration with standard hydrochloric acid solution

E   Titration with standard iodine solution

18. The purpose of the sodium thiosulphate is to

A   react with some iodine

B   react with some potassium iodide

C   react with some peroxydisulphate

D   catalyse the overall reaction

E   act as an oxidising agent

19. A rate equation which would be consistent with the given data would be

A   rate $\propto [S_2 O_8^{2-}]$

B   rate $\propto [I^-]$

C   rate $\propto [I^-][S_2 O_8^{2-}]$

D   rate $\propto [I^-]^2 [S_2 O_8^{2-}]$

E   rate $\propto [I^-][S_2 O_8^{2-}]^2$

20. In a further experimental run, the initial concentrations were:

peroxydisulphate 0.10 M
iodide 0.15 M

The expected time, in seconds, for the appearance of the dark colour would be

A 36

B 47

C 71

D 87

E 105

Questions 21–24 concern an investigation of the kinetics of the acid-catalysed iodination of acetone. This reaction is first order with respect to hydrogen ions

$$CH_3COCH_3 + I_2 \xrightarrow{H^+} CH_3COCH_2I + HI$$

The following results were obtained for the time required to reduce the initial iodine concentration by a constant amount. The hydrogen ion concentration was maintained constant throughout.

$[CH_3COCH_3]$/mol dm^{-3}	$[I_2]$/mol dm^{-3}	*Time*/min
0.25	0.050	7.2
0.50	0.050	3.6
1.0	0.050	1.8
0.50	0.10	3.6

21. The order of reaction with respect to acetone is

A 0

B 1

C 2

D 3

E more complex

22. The order of reaction with respect to iodine is

A 0

B 1

C 2

D 3

E more complex

23. A possible expression for the rate of reaction (k denotes the rate constant) is

A $k[CH_3COCH_3][I_2]$

B $k[CH_3COCH_3][H^+]$

C $k[CH_3COCH_3][I_2][H^+]$

D $k[CH_3COCH_3]^2[H^+]$

E more complex

24. A possible mechanism for the reaction is

A $CH_3COCH_3 + I_2 \xrightarrow{H^+} CH_3COCH_2I + HI$

B $CH_3COCH_3 \xrightarrow{H^+} CH_3\underset{\underset{OH}{|}}{C} = CH_2$ (fast)

$CH_3\underset{\underset{OH}{|}}{C} = CH_2 + I_2 \longrightarrow CH_3COCH_2I + HI$ (slow)

C $CH_3COCH_3 \xrightarrow{H^+} CH_3\underset{\underset{OH}{|}}{C} = CH_2$ (slow)

$CH_3\underset{\underset{OH}{|}}{C} = CH_2 + I_2 \longrightarrow CH_3COCH_2I + HI$ (fast)

D $CH_3COCH_3 \longrightarrow CH_3COCH_2^- + H^+$ (slow)

$I_2 \longrightarrow I^+ + I^-$ (slow)

$CH_3COCH_2^- + I^+ \longrightarrow CH_3COCH_2I$ (fast)

E $CH_3COCH_3 \longrightarrow CH_3COCH_2^- + H^+$ (fast)

$I_2 \longrightarrow I^+ + I^-$ (slow)

$CH_3COCH_2^- + I^+ \longrightarrow CH_3COCH_2I$ (fast)

Test 8   Answers

1. The concentration of HI at the beginning of the reaction ($X = 0$) will be zero. Only curves **A** and **D** are possible. The reaction between H_2 and I_2 will start rapidly and will gradually slow down as the concentrations of H_2 and I_2 decrease. The rate of increase in the HI concentration will therefore decrease and eventually this concentration will reach a limiting value. This is shown by curve **A**, the correct key.

   There is another reason for the HI concentration reaching a limiting value in this reaction. Do you know what this second reason is? Refer, if necessary, to test 9, question 21 for a hint about the answer.

2. The concentration of ethyl acetate will have its highest value at the time of mixing and will decrease from then on. Only curves **C** and **E** are possible. Curve **C** suggests that the reaction is zero order with respect to both reactants, which appears to be most unlikely. Curve **E** shows the rate of reaction decreasing as the concentrations of reactants decrease, which is what is to be expected. It also appears in **E** that the concentration of ethyl acetate is only approaching a limiting value of zero after an infinite time which is what would be expected for a first or second order hydrolysis that went to completion [*see* question 6 which gives the order of a similar reaction]. The correct key is **E**.

3. The rate equation will be

   Rate = rate constant $\times$ (concentration of reactant)^0

   or Y = rate constant $\times X^0$

   or Y = rate constant (as X^0 is 1)

   This is curve **B**.

4. Bromine is a reactant rather than a catalyst (**3** incorrect). Although the reaction rate does not depend on the bromine concentration, it will depend on other concentrations which will alter during the reaction (**1** incorrect).

   The zero order of this reaction with respect to bromine shows that the slow, or rate-determining, step does not involve bromine. **2** is incorrect and, with only **4** correct, the key is **D**.

5. The initiation reaction involves bond breaking, an endothermic reaction, so **1** is incorrect.

   In stage **II**, a C—H bond is broken (409 kJ mol^{-1} absorbed), while an H—Br bond is formed (367 kJ mol^{-1} emitted). 42 kJ mol^{-1} will be absorbed at this stage and **2** is incorrect.

   In stage **III**, a Br—Br bond is broken (192 kJ mol^{-1} absorbed) and a C—Br bond is formed (284 kJ mol^{-1} emitted). 92 kJ mol^{-1} will be emitted at this stage and **3** is incorrect.

   42 kJ mol^{-1} is absorbed in stage **II** while 92 kJ mol^{-1} is emitted in stage **III**. Stages **II** and **III** combined are exothermic. **4** is the only correct statement and the key is **D**.

6. The information shows that

   Rate = k[ethyl propanoate]^1[OH$^-$]^1

   1 and **3** are the only true statements that follow from this. The key is **B**. The reaction is second order overall (**2** incorrect). Halving the concentrations of both reactants simultaneously will quarter the rate of reaction (**4** incorrect).

7. This question is asking which statements follow correctly from the given information. The reaction in the reverse direction must take the same route, in reverse, as the forward reaction. The activation energy of the reverse reaction is $(x + y)$ so **1** follows correctly. This is the case because the reverse reaction would not take a route with a higher activation energy than necessary and there cannot be a route with lower activation energy or the forward reaction would have found, and used, this other route.

   The forward reaction is exothermic, with energy change y. **3** follows correctly. ΔH for the reverse reaction also has the magnitude y (endothermic) so **2** is incorrect. The chemical equation gives no information about the order or mechanism of a reaction and hence it cannot be said that **4** follows from the information given. Only **1** and **3** follow correctly and the key is **B**.

8. At low urea concentrations, the graph shows an increase in initial rate with increasing urea concentration. The initial rate depends on urea concentration and **1** is correct.

At the highest concentrations covered, the horizontal nature of the curve in the graph shows that there is no change in initial rate with increasing urea concentration. The initial rate is independent of urea concentration (**2** correct). This is another way of saying that the reaction is zero order with respect to urea (**3** correct).

The experiment is concerned only with the rate at the start of the reaction. There is no information about what happens as hydrolysis proceeds. **4** could not be regarded as a suggestion that followed from the results given. If an attempt is made to predict what will happen to the rate as hydrolysis proceeds, the conclusion is that the rate will either decrease, if the urea concentration is low, or, if the urea concentration is sufficiently high, the rate will remain steady for a while and then decrease when sufficient urea has been consumed. **4** is still incorrect and the key is **A**.

9. All reactants and products are in aqueous solution and any volume change will be negligible; if the substances were pure liquids or if a gas was evolved, this method is likely to be satisfactory. All substances are too simple in structure to possess the lack of symmetry on which optical activity depends. **1** and **3** are incorrect.

The reaction produces coloured iodine from colourless reactants while ions become molecules. The colour intensity and electrical conductivity of the system will alter considerably during the reaction and these properties can be used to obtain information about the rate. **2** and **4** are correct and the key is **C**.

10. The rate will increase by a factor of 4^2, which is 16, owing to the change in concentration of x. The change in the concentration of y will increase the rate by a factor of $4^{1/2}$, which is $\sqrt{4}$ or 2. The combined (multiplied) effect of these

two changes in concentration will be to increase the rate by a factor of 16×2, or 32, which is key **D**.

11. The rate equation for the reaction is

$$\text{Rate} = k\,[\text{reactant}]^1$$

As time passes and the reaction proceeds, the concentration of the reactant decreases and so does the rate. Only **C** and **E** show a decrease in rate with time and choice has to be made between these two.

C indicates that the rate will become zero after a finite time and this makes it incorrect. For a first order reaction, the concentration of reactant decreases in an exponential way with time and this means that the reaction will never really finish, only become slower and slower. This is shown in **E** which is the correct key. The curve in **E** is rather similar to the concentration–time curve for a first order reaction and it is possible that you thought that **E** was the concentration–time curve and so chose the correct key for the wrong reason!

12. The rate equation given in the question suggests that the slow, rate-determining step in the mechanism involves only one molecule of X(aq) and one OH^-(aq) ion. Y must take its part in the mechanism in a fast step. The only mechanism showing a slow step that is consistent with the rate equation is **E** and this also has Y in the fast step.

13. Note the wording of the question. Four of the rate equations could be correct. You are looking for the one that must be INCORRECT. The rate equation is of the form

$$\text{Rate} = k\,[\text{X}]^a\,[\text{Y}]^b\,[\text{others}]^c$$

The overall reaction order of 3 shows that $(a + b + c)$ is 3. This is the case with **B**, **C**, **D** and **E** but not for **A** which is the key in this question.

14. The ionic equation for the reaction being studied is

$$2MnO_4^-(aq) + 5C_2O_4^{2-}(aq) + 16H^+(aq) \rightarrow$$
$$2Mn^{2+}(aq) + 10CO_2(g) + 8H_2O(l)$$

The reaction is catalysed by a product, $Mn^{2+}(aq)$, and proceeds very slowly until sufficient catalyst has accumulated to exert a strong influence on the rate. After an initial waiting period, the reaction rate suddenly accelerates as more reaction produces more catalyst which makes the reaction go even faster. In the description of the experiment, the reaction appears to increase in rate as more permanganate is added. This may really be because it takes time to add more permanganate and time is what the reaction needs in order to produce a trace of catalyst. If the permanganate is added slowly enough, the increase in rate could be because of the increasing concentration of catalyst produced.

The correct key is **A** and this is an easy question if you already know the answer! If you had not met the reaction before, you could certainly eliminate **B** because, although true, an increase in pH, and hence decrease in $H^+(aq)$ concentration, would slow down the reaction. You could also eliminate **E** because the sulphuric acid is not concentrated. **C** sounds plausible but have you ever met a reaction where this is true? **D** is probably the best wrong answer but the reaction would have to be very exothermic and the solutions of high concentration for the temperature rise to produce a marked increase in rate. Of course, if you have not met the reaction before you might well select **D** because you are taking a chance on the reaction being a highly exothermic one between concentrated solutions. In this reaction, the temperature has to approach about 60 °C before there is a really significant increase in rate.

15. The time halves, and hence rate doubles, when the concentration of Y is doubled between the first and second set of readings. The reaction must be first order with respect to Y.

    Between the first and third set of figures, the concentration of X has doubled but there has been no effect on the time and hence rate of reaction. The reaction must be zero order with respect to X. The two orders are shown correctly in **B**.

16. The reaction works by peroxydisulphate ions oxidizing iodide ions to iodine molecules. The iodine is at first immediately consumed by the thiosulphate that is also present. Suddenly,

when the thiosulphate has all been used up, the further iodine produced turns starch to the familiar dark blue colour. The key is **C**. You need to recognize that starch is being called a polysaccharide.

17. A colorimeter could be used to detect the sudden appearance of the starch—iodine complex, key **B**. There would be no need to carry out preliminary experiments to find the best filter to use in the colorimeter because all that is wanted is a measurement of the one time when sudden, intense absorption begins.

18. The purpose of the thiosulphate is, as mentioned earlier, to consume the iodine produced at the start of the experiment, key **A**. The volume and molarity of the thiosulphate show that 5×10^{-5} moles of thiosulphate are used which means that the experiment is really measuring the time taken to produce 2.5×10^{-5} moles of iodine. The equation for the reaction between peroxydisulphate and iodide is

$$S_2O_8^{2-} + 2I^- \rightarrow 2SO_4^{2-} + I_2$$

Look at the table and decide for the first reaction how many moles of iodine will be formed when the reaction is complete. Do you think it is correct to describe this experiment as one 'to determine the *initial* rate of reaction'? (To answer this question you need to use the volumes and initial concentrations of $(NH_4)_2S_2O_8$ and KI. You also need to use the equation for the reaction between $S_2O_8^{2-}$ and I^- given above.)

19. Inspection of the first, fourth and fifth experiments in the table shows that as the iodide concentration is reduced to one half or one third of its original value, the time increases by a factor of 2 or 3. This means that the rate is reduced to one half or one third of its original value. The order of this reaction is unity with respect to iodide ions. A very similar reduction in rate is found for a similar, but not identical, reduction in peroxydisulphate concentration — this comes from inspection of the first three experiments in the table. The reaction appears to be first order with respect to peroxydisulphate. The first order dependence on both concentrations is covered by key **C**.

20. The peroxydisulphate concentration is the same as in the first and fourth experiments while the iodide concentration is midway between its values for those experiments in the table. For this reason, a time approximately mid-way between 35 and 70 seconds would be expected and only one time, key **B**, lies in this range. How close is the 47 seconds of key **B** to what would be expected? You could check this by looking at the first set of values in the table and deciding how the time of 35 seconds would be affected by a change in iodide molarity from 0·2 to 0·15.

21. Taking the first two sets of figures in the table, as the acetone concentration is doubled the time is halved and hence the rate doubles. The order with respect to acetone is unity, key **B**.

22. Comparing the second and fourth sets of figures in the table, as the iodine concentration is doubled the time and reaction rate remain unchanged. The order with respect to iodine is zero, key **A**.

23. The answers to questions 21 and 22 and the information at the start about the reactions being first order with respect to hydrogen ions suggest the rate equation given in key **B**.

24. The rate equation indicates that only acetone molecules and hydrogen ions are involved in the slow, rate-determining step in the reaction. The only suggestion made for a mechanism which is possible is that given in key **C**.

Test 9

Equilibrium 1 (Gaseous, acid–base and solubility)

Questions 1–3 concern the following graphs which show the change in pH of 20 cm³ of 1M alkali solution when 1M acid solution is added.

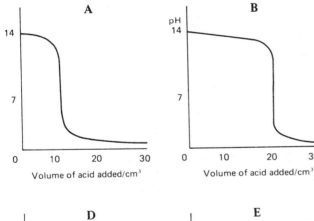

A

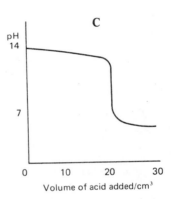

B

C

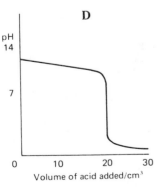

D

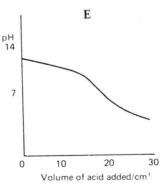

E

Select the graph which indicates the change occurring for the addition of

1. propanoic acid to potassium hydroxide

2. sulphuric acid to sodium hydroxide

3. nitric acid to ammonia solution

Directions summarized for questions 4 to 10

A	B	C	D	E
1,2,3 only correct	1,3 only correct	2,4 only correct	4 only correct	Some other response or combination of responses is correct

4 Ethanoic acid (acetic acid) is a weak acid and hydrochloric acid is a strong acid.

It follows that the

1 pH of 0·1M hydrochloric acid will be approximately 1

2 solution containing 0.1 M ethanoic acid and 0.1 M sodium ethanoate (sodium acetate) will be a good buffer

3 pH of 0·1M hydrochloric acid will be less than that of 0·1M ethanoic acid

4 pH of a solution formed by mixing equimolar quantities of sodium hydroxide and hydrochloric acid will be greater than that of a similar solution formed from sodium hydroxide and ethanoic acid

5. Mixtures which, when dissolved in water, act as buffer solutions include

1 HNO_3 and $NaNO_3$

2 NH_3 and NH_4Cl

3 CH_3CO_2H and $NaCl$

4 NaH_2PO_4 and Na_2HPO_4

6. For the reaction $N_2O_4(g) \rightleftharpoons 2NO_2(g)$

K_p at 600 °C is $1·78 \times 10^4$

and K_p at 1000 °C is $2·82 \times 10^4$

It follows that

1 ΔH is negative

2 the formation of NO_2 is favoured by rise in temperature

3 the units of K_p are $dm^3 \, atm^{-1}$

4 the formation of NO_2 is favoured by decrease in pressure

7. The reaction

$$2BaO_2(s) \rightleftharpoons 2BaO(s) + O_2(g)$$

was studied by heating, at a fixed temperature, a sample of BaO_2 in an evacuated glass vessel connected to a **U**-tube manometer. For this system in equilibrium, it would be found that

1 the pressure of oxygen is proportional to the mass of BaO_2

2 the pressure of oxygen is equal to the equilibrium constant K_p

3 the pressure of oxygen is inversely proportional to the mass of BaO formed

4 if oxygen is introduced into the system, it reacts with the BaO until the original pressure of oxygen at that temperature is again reached

8. If a solution which is 0·0001M with respect to carbonate ions, CO_3^{2-}, is mixed with an equal volume of a 0·0001M solution of ions of a Group II metal, which of the following carbonates would be precipitated?

$K_{sp}(298 \text{ K})/mol^2 \, dm^{-6}$

1 $MgCO_3$ $1·1 \times 10^{-5}$

2 $CaCO_3$ $5·0 \times 10^{-9}$

3 $SrCO_3$ $1·1 \times 10^{-10}$

4 $BaCO_3$ $5·5 \times 10^{-10}$

9. The dissociation constants K_a are given for a group of acids:

	K_a/mol dm^{-3}
Phenylethanoic acid	
($C_6H_5CH_2CO_2H$)	5.5×10^{-5}
3-Hydroxybenzoic acid	8.3×10^{-5}
2-Chlorobutanoic acid	1.4×10^{-3}

Correct statements about these three acids include that

1 K_a for phenylethanoic acid is given by

$$\frac{[C_6H_5CH_2CO_2^-][H^+]}{[C_6H_5CH_2CO_2H]}$$

2 $[H^+]$ in 1.0M 3-hydroxybenzoic acid solution is 8.3×10^{-5} mol dm^{-3}

3 K_a for 2-chlorobutanoic acid is less than K_a for 2,2-dichlorobutanoic acid

4 3-hydroxybenzoic acid is the strongest acid

10. Values of K_w are

0.64×10^{-14} mol^2 dm^{-6} at 18 °C

1.00×10^{-14} mol^2 dm^{-6} at 25 °C

From this it may be deduced that

1 the ionization of water is an endothermic process

2 the pH of water is greater at 25 °C than at 18 °C

3 the hydroxyl ion concentration in water at 18 °C is 0.8×10^{-7} mol dm^{-3}

4 water is only neutral at 25 °C

Directions for questions 11 to 20. Each of the questions or incomplete statements in this section is followed by five suggested answers. Select the best answer in each case.

11. At a certain temperature, for the reaction

$$CO(g) + Cl_2(g) \rightleftharpoons COCl_2(g)$$

the equilibrium partial pressures of carbon monoxide, chlorine and carbonyl chloride are 2, 4 and 48 atm, respectively. What is the value of K_p?

A 54 atm

B 42 atm

C 12 atm

D 8 atm^{-1}

E 6 atm^{-1}

12. At a certain temperature the equilibrium constant for the reaction

$$CO(g) + H_2O(g) \rightleftharpoons CO_2(g) + H_2(g)$$

is 4. A mixture initially containing one mole of each of carbon monoxide and steam is allowed to reach equilibrium. How many moles of carbon monoxide are now present?

A $\frac{1}{4}$

B $\frac{1}{3}$

C $\frac{1}{2}$

D $\frac{2}{3}$

E $\frac{3}{4}$

13. The solubility product of silver chromate is 1×10^{-12} mol^3 dm^{-9}. In a solution in which $[CrO_4^{2-}]$ is 1×10^{-4} mol dm^{-3}, the maximum $[Ag^+]$, in mol dm^{-3}, is

A 0.5×10^{-8}

B 1×10^{-8}

C 1×10^{-6}

D 0.5×10^{-4}

E 1×10^{-4}

14. The solubility product for mercury(II) sulphide in aqueous solution at 25 °C is given in data books as 1.60×10^{-52} mol^2 dm^{-6}. Suppose the substance exists in solution as Hg^{2+}(aq) and S^{2-}(aq) ions, of what order of magnitude is the number of individual Hg^{2+}(aq) ions per dm^3 of saturated solution at 25 °C?

(Avogadro constant $= 6 \times 10^{23}$ mol^{-1})

A 10^{-28}

B 10^{-26}

C 10^{-2}

D 10^2

E 10^{26}

15. A sparingly soluble iron(III) salt dissociates in solution thus:

$$Fe_2 X_3 (s) \rightleftharpoons 2Fe^{3+} (aq) + 3X^{2-}(aq)$$

If the solubility product for the iron(III) salt is K_{sp}, then $[Fe^{3+}(aq)]_{eqm}$ is

A $(K_{sp})^{1/2}$

B $(\frac{2}{3}K_{sp})^{1/2}$

C $(\frac{8}{27}K_{sp})^{1/5}$

D $(\frac{2}{3}K_{sp})^{1/3}$

E $(\frac{8}{27}K_{sp})^{1/3}$

16. The dissociation constant K_a of a weak monobasic acid is 10^{-7} mol dm^{-3}. What is the approximate pH of a 0.1 M solution of the acid?

A 1

B 2

C 3

D 4

E 5

17. What is the approximate pH of a buffer solution containing 0.20 mole of a weak monobasic acid ($K_a = 10^{-4.8}$ mol dm^{-3}) and 0.02 mole of the sodium salt of the acid in 1 dm^3 of aqueous solution?

A 2.8

B 3.8

C 4.8

D 5.8

E 6.8

18. The ionization constant for ammonia is given by the expression

$$\frac{[NH_4^+(aq)]\,[OH^-(aq)]}{[NH_3(aq)]}$$

and is approximately 10^{-5} mol dm^{-3} at 25 °C. K_w is 10^{-14} mol^2 dm^{-6} at 25 °C.

A mixture of equal volumes of 1 M ammonia and 1 M ammonium chloride will have a pH most nearly equal to

A 3

B 5

C 7

D 9

E 11

19. The acid dissociation constant for the reaction

$$C_6H_5NH_3^+(aq) \rightleftharpoons C_6H_5NH_2(aq) + H^+(aq)$$

is

$$K_a = \frac{[C_6H_5NH_2(aq)] \, [H^+(aq)]}{[C_6H_5NH_3^+(aq)]}$$

If α denotes the degree of dissociation and x mol dm^{-3} the initial concentration of $C_6H_5NH_3^+(aq)$, the value of K_a is

A $\dfrac{x\alpha^2}{1-\alpha}$

B $\dfrac{\alpha^2}{x(1-\alpha)}$

C $\dfrac{x\alpha^2}{1+\alpha}$

D $\dfrac{\alpha^2}{x(1+\alpha)}$

E $\dfrac{x\alpha^2}{(1-\alpha)^2}$

20. Which diagram could represent the variation in the concentrations of X and Y with time in the reversible reaction $X \rightleftharpoons Y$ which comes to equilibrium after a time t?

A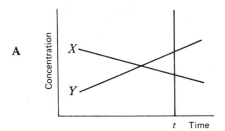

B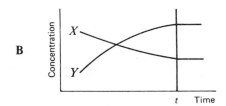

C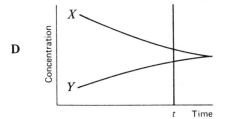

D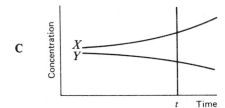

E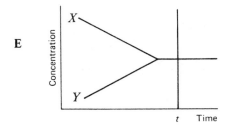

21. 28 g of nitrogen (N = 14) and 6 g of hydrogen (H = 1) are enclosed in an inert metal cylinder of fixed volume and heated until equilibrium is reached at 450 °C.

$$N_2 + 3H_2 \rightleftharpoons 2NH_3 ; \Delta H = -101 \text{ kJ}$$

Which one of the following changes would yield the greatest mass of ammonia in the cylinder when equilibrium was re-established?

A Addition of 6 g of hydrogen without change in temperature

B Addition of 28 g of nitrogen without change in temperature

C Addition of 6 g of hydrogen with cooling to 350 °C

D Addition of 28 g of nitrogen with cooling to 350 °C

E Addition of finely divided iron as a catalyst with cooling to 350 °C

22. An equilibrium mixture contains nitrogen, hydrogen and ammonia at the following partial pressures

N_2 100; H_2 10; NH_3 50 (arbitrary units).

What value fo these figures give for the equilibrium constant, K_p, for the reaction

$$2NH_3 \rightleftharpoons N_2 + 3H_2 ?$$

A $\dfrac{1}{40}$ D 30

B $\dfrac{1}{20}$ E 40

C 20

23 For the reaction

$$CH_3 CO_2 C_2 H_5 (1) + H_2 O(1) \rightleftharpoons$$
$$CH_3 CO_2 H(1) + C_2 H_5 OH(1)$$

the equilibrium constant, K_c at 298 K is ¼.

If 1 mole of ester and 1 mole of water were mixed and allowed to come to equilibrium, the number of moles of ethanoic (acetic) acid formed would be

A ¼   C ⅜   D ⅔

B ⅓   E ¾

24. Calcium carbonate, $CaCO_3$, precipitates from a 0.2 M solution of calcium nitrate when the concentration of carbonate ion reaches 2.5×10^{-8} mol dm^{-3} What is the solubility product of calcium carbonate in mol^2 dm^{-6}?

A 1.25×10^{-16} C 5×10^{-9} E 5×10^{-7}

B 1×10^{-9} D 1.25×10^{-7}

25. The hydrogen ion concentration, in mol dm^{-3}, in 0.2 M ethanoic acid (K_a is 2×10^{-5} mol dm^{-3}) is approximately

A 2×10^{-2} D 2×10^{-5}

B 2×10^{-3} E 4×10^{-6}

C 4×10^{-4}

26. What is the pH of the buffer solution formed by mixing equal volumes of solutions of 0.1 M propanoic acid and 0.1 M sodium propanoate? (The dissociation constant for propanoic acid is 1.3×10^{-5} mol dm^{-3})

A 1.00   D 4.89

B 1.30   E 5.11

C 3.11

27. The hydrogen ion concentration of a buffer solution, consisting of a weak acid in an aqueous solution of its sodium salt, is given by

A $[H^+] = K_a [\text{Salt}]$ D $[H^+] = K_a \dfrac{[\text{Acid}]}{[\text{Salt}]}$

B $[H^+] = K_a [\text{Acid}]$ E $[H^+] = K_a \dfrac{[\text{Salt}]}{[\text{Acid}]}$

C $[H^+] = K_a [\text{Acid} + \text{Salt}]$

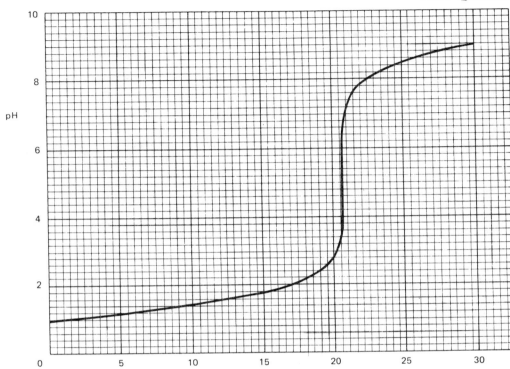

Volume of base added/cm³

Questions 28–30

A solution of a base was added from a burette to 25 cm³ of acid and the pH of the mixture noted at intervals during the addition. The graph shows the changes in pH during addition of base to 25 cm³ of acid.

28. Which of the following was the acid used?

A $0.1 M\ CH_3CO_2H$

B $1 M\ CH_3CO_2H$

C $0.1 M\ HCl$

D $1 M\ HCl$

E $1 M\ H_2SO_4$

29. If the acid solution was x M with respect to H^+ ions, what was the apparent molarity of the basic solution with respect to OH^- ions?

A $0.80x$ M

B $0.82x$ M

C $1.00x$ M

D $1.22x$ M

E $1.25x$ M

30. Which of the following was the most likely basic solution to be used in the experiment?

A Aqueous ammonia

B Barium hydroxide

C Lithium hydroxide

D Sodium hydroxide

E Potassium hydroxide

Test 9 Answers

1. Potassium hydroxide is a strongly ionized base. In a 1 M solution, the concentration of hydroxide ions will be 1 mole dm^{-3} and this will make the concentration of hydrogen ions take a value of 10^{-14} mol dm^{-3} (see the answer to question 10). The pH will start at 14 and so choice has to be made between **A**, **B** and **C**. Propanoic acid is a weakly ionized acid and the pH of the mixture after the end point will not tend to a very low value. The curve in **C** is the only possible one.

2. Sodium hydroxide is another strongly ionized base so, as in question 1, the only possible curves are those given in **A**, **B** and **C**. An equation for the reaction taking place during this titration is

$$2NaOH + H_2SO_4 \rightarrow Na_2SO_4 + 2H_2O$$

Every mole of sodium hydroxide will require only half a mole of sulphuric acid for neutralization. 20 cm^3 of 1 M alkali will require only 10 cm^3 of the same molarity acid. The correct curve must be **A**.

3. The ionization of ammonia solution to produce hydroxide ions in this solution will be considerably less than 1 mole dm^{-3}. The concentration of hydrogen ions will be more than 10^{-14} mol dm^{-3} and the pH will be less than 14. The only possible curves are those in **D** and **E**. Nitric acid is a strongly ionized acid which reacts in the ratio 1 mole to 1 mole with ammonia solution. The curve must be that in **D**.

   If a very large volume of nitric acid is added before the titration is stopped, the solution would approximate to 1 M nitric acid and the pH would be 0. The solution in the first question would be 1 M propanoic acid when titration had been continued well beyond the end-point. In question 2, the final solution would tend to 1 M sulphuric acid. Suggest rough values for the limiting value of the pH in these two cases.

4. Because of complete ionization, 0·1 M HCl will produce a hydrogen ion concentration of

0·1 mole dm^{-3} which is 10^{-1} mole dm^{-3}. The pH will be 1 and the first statement is correct. Can you suggest why the word 'approximately' is used?

In a solution containing ethanoic acid and sodium ethanoate the following equilibrium exists

$$CH_3CO_2H + H_2O \rightleftharpoons H_3O^+ + CH_3CO_2^-$$

If hydrogen ions H$_3$O$^+$ are removed from the solution, there is a good supply of ethanoic acid to ionize and produce more hydrogen ions to restore the disturbed equilibrium. There is also a good supply of ethanoate ions and these are able to combine with added hydrogen ions when the equilibrium position is disturbed by addition of acid. The solution will resist attempts to change its hydrogen ion concentration and would be described as a good buffer. 2 is correct.

Because of incomplete ionization, the concentration of hydrogen ions in ethanoic acid will be lower than in a solution of HCl of equal molarity. The pH of the HCl will be lower and 3 is correct.

One way to decide on the fourth statement is to regard these mixtures as being those produced at the end points in titrations. The first mixture will have a pH close to 7 as it is the result of titrating the strong acid HCl against the strong base NaOH. The titration curve will be similar to **B** in questions 1 to 3. The second mixture will have a pH above 7 as it is the result of titrating the weak acid ethanoic against the strong base NaOH. The titration curve will be similar to **C** in questions 1 to 3. The fourth statement is not correct and the key is **A**.

The other way to decide on the fourth statement is to realise that these reactions will have produced aqueous solutions of the salts sodium chloride and sodium ethanoate. Sodium chloride is not hydrolysed and its pH is close to 7. Sodium ethanoate is the salt of a weak acid and strong base and will be alkaline by hydrolysis.

$$CH_3CO_2^- + H_2O \rightleftharpoons CH_3CO_2H + OH^-$$

5. A buffer solution is frequently made up from a weak acid and one of its salts, or from a weak base and one of its salts. As with the mixture of ethanoic acid and sodium ethanoate in question 4, this gives the mixture the ability to cope with both removal and addition of hydrogen ions (*see* the second paragraph of the answer to question 4).

Mixtures 2 and 4 are of this type and the key is **C**. The equilibria present in these mixtures are shown below.

$$NH_3 + H_3O^+ \rightleftharpoons NH_4^+ + H_2O$$
$$H_2PO_4^- + H_2O \rightleftharpoons H_3O^+ + HPO_4^{2-}$$

6. This equilibrium constant increases with increasing temperature. Reaction moves to the right, forming more NO_2 (2 correct). This must be the endothermic direction (1 incorrect). Where P_e stands for the equilibrium partial pressure,

$$K_p = \frac{P_e^2(NO_2)}{P_e(N_2O_4)} \text{ so units must be } \frac{atm^2}{atm} = atm$$

The third statement is incorrect as the units should be atm.

Decrease in pressure favours the reaction direction which produces the greater number of gaseous molecules. The formation of NO_2 is favoured and 4 is correct. The key is **C**.

7. The solids in this equilibrium do not appear in the equilibrium constant expression which is

$$K_p = P_e O_2$$

The oxygen pressure at equilibrium $P_e(O_2)$ is equal to the equilibrium constant (2 correct). Temperature change is the only factor that can alter an equilibrium constant and therefore the oxygen pressure is unaffected by the changes suggested in 1 and 3 which are incorrect.

4 is correct because as long as BaO remains it will absorb added oxygen until the equilibrium pressure is regained. The key is **C**.

8. The solubility equilibria are of the type

$$MCO_3(s) + aq \rightleftharpoons M^{2+}(aq) + CO_3^{2-}(aq)$$

The equilibrium constant expression is

$$K_{sp} = [M^{2+}(aq)]_e [CO_3^{2-}(aq)]_e$$

Note the 'e' by each concentration indicating that it is an equilibrium concentration. It is only at equilibrium that the product of the concentrations is equal to K_{sp}. If the product of the concentrations exceeds K_{sp}, a precipitate will begin to form and precipitation will continue until the product is equal to K_{sp}.

On mixing the solutions in the problem, each dilutes the other and the concentrations of metal ions and carbonate ions both become 5×10^{-5} mol dm^{-3}. The product of these concentrations is 2.5×10^{-9} mol^2 dm^{-6}. This exceeds the K_{sp} values for the carbonates of Ca, Sr and Ba. These carbonates will precipitate until each K_{sp} value is reached. The key is **E**.

9. The dissociation of phenylethanoic acid in aqueous solution can be written

$$C_6H_5CH_2CO_2H(aq) \rightleftharpoons$$
$$C_6H_5CH_2CO_2^-(aq) + H^+(aq)$$

The dissociation constant connects the concentrations of the species in the reaction at the equilibrium

$$K_a = \frac{[C_6H_5CH_2CO_2^-(aq)]_e [H^+(aq)]_e}{[C_6H_5CH_2CO_2H(aq)]_e}$$

The first statement in the question is the same as this except that state symbols have been omitted and it is not completely clear that the concentrations really are equilibrium ones. 1 is meant to be correct (but see comment at end).

If the equilibrium concentration of $H^+(aq)$ in 3-hydroxybenzoic acid is taken to be x, the concentration of the anion of the acid will also be x. The concentration of un-ionized acid will still be close to 1 M as so little will be ionized. Putting these values into an equilibrium constant similar to that in the first statement gives $(x \times x)/1 = 8.3 \times 10^{-5}$, whence $x = \sqrt{(8.3 \times 10^{-5})}$ mol dm^{-3}. The concentration of $H^+(aq)$ is not that given in 2.

The dissociation of organic acids is increased by the presence of halogen atoms in the molecule, particularly if the halogen is attached on the carbon atom next to the CO_2H group. Two halogen atoms are more effective than one. This makes **3** correct. The strongest acid will have the highest dissociation constant and this acid is 2-chlorobutanoic which makes **4** incorrect.

1 and **3** are correct and the key is **B**. [If you are required on your course to include state symbols in equilibrium constants and to indicate that concentrations are equilibrium values, you can take the key to be **E**.]

10. Water becomes slightly more ionized on raising the temperature so ionization must be endothermic. **1** is correct. Because water is more highly ionized at 25 °C, the equilibrium concentration of $H^+(aq)$ will be higher at this temperature and the pH will be lower (**2** incorrect).

At 18 °C,

$$[H^+(aq)]_e[OH^-(aq)]_e = 0.64 \times 10^{-14} \, mol^2 \, dm^{-6}$$

The hydrogen and hydroxyl ion concentrations are equal and the hydroxyl ion concentration will have the value and units given in **3**.

Water is neutral if it contains equal concentrations of hydrogen and hydroxyl ions. This will be true for neutral water at all temperatures. The concentrations of the two ions will alter as temperature changes but will remain equal. Only **1** and **3** are correct and the key is **B**.

11.
$$K_p = \frac{P_e(COCl_2)}{P_e(CO) \times P_e(Cl_2)} =$$

$$\frac{48 \, atm}{2 \, atm \times 4 \, atm} = 6 \, atm^{-1} \text{ (key E)}.$$

12.
$$K_p = \frac{P_e(CO_2) \times P_e(H_2)}{P_e(CO) \times P_e(H_2O)} = 4$$

Note that K_p has no units. This is the situation where K_p and K_c have the same value and where the equilibrium position is not dependent on volume or pressure. Take the volume used in the problem to be the molar volume:

At the start $P(CO) = P(H_2O) = 1 \, atm$, $P(CO_2) = P(H_2) = 0$

At equilibrium $P_e(CO) = P_e(H_2O) = (1 - x)$ atm, $P_e(CO_2) = P_e(H_2) = x$ atm

Putting these values into the equilibrium constant gives

$$\frac{x^2}{(1-x)^2} = 4$$

Taking square roots of both sides

$$\frac{x}{1-x} = +2 \text{ (or } -2 \text{ but this does not give a possible solution)}$$

Solving for x gives a value of 2/3 so the number of moles of CO remaining will be 1/3, key **B**.

13. $[Ag^+(aq)]_e^2[CrO_4^{2-}(aq)]_e$
$$= K_{sp}$$
$$= 1 \times 10^{-12} \, mol^3 \, dm^{-9}$$

$$[Ag^+(aq)]_e^2 \times 1 \times 10^{-4} \, mol \, dm^{-3} = 1 \times 10^{-12} \, mol^3 \, dm^{-9}$$

Solving this equation gives an equilibrium concentration of silver ions of $10^{-4} \, mol \, dm^{-3}$. This is the maximum value the concentration can have so the key is **E**.

14. $[Hg^{2+}(aq)]_e[S^{2-}(aq)]_e$
$$= 1.6 \times 10^{-52} \, mol^2 \, dm^{-6}$$

In a saturated solution, the concentrations of mercury(II) and sulphide ions are the equilibrium values and equal

$$[Hg^{2+}(aq)]_e = \sqrt{(1.6 \times 10^{-52})} \, mol \, dm^{-3}$$

which is about $1.3 \times 10^{-26} \, mol \, dm^{-3}$. The number of $Hg^{2+}(aq)$ ions will be

$$1.3 \times 10^{-26} \, mol \, dm^{-3} \times 6 \times 10^{23} \, mol^{-1}$$

This number is about $8 \times 10^{-3} \, dm^{-3}$ which, as an order of magnitude, is $10^{-2} \, dm^{-3}$, key **C**.

Mercury(II) sulphide is one of the most insoluble salts known. This approximate calculation shows that in a saturated solution of the salt there is only one $Hg^{2+}(aq)$ ion in every 100 dm^3.

15. $[Fe^{3+}(aq)]_e^2[X^{2-}(aq)]_e^3 = K_{sp}$

If the equilibrium concentration of $Fe^{3+}(aq)$ is taken as x, the equilibrium concentration of $X^{2-}(aq)$ will be $(3/2)x$.

$$x^2 \left(\frac{3x}{2}\right)^3 = \frac{27x^5}{8} = K_{sp}$$

$$x = \left(\frac{8}{27} K_{sp}\right)^{1/5} \text{ which is key C.}$$

16. If the acid is given the formula HA,

$$HA(aq) \rightleftharpoons H^+(aq) + A^-(aq)$$

$$K_a = \frac{[H^+(aq)]_e [A^-(aq)]_e}{[HA(aq)]_e} = 10^{-7} \text{ mol dm}^{-3}$$

If $[H^+(aq)]_e$ is given the value x, this will also be the concentration of A^- while the concentration of HA will be $(0.1 - x)$ [all in concentration units]. Substitution in the equilibrium constant expression gives

$$\frac{x^2}{0.1 - x} = 10^{-7}$$

The acid is weakly ionized and x can be neglected in comparison with 0.1. The equation becomes

$$x^2/0.1 = 10^{-7}; \quad x^2 = 10^{-8}; \quad x = 10^{-4}$$

The hydrogen ion concentration is 10^{-4} giving a pH of 4, key **D**. You may find it difficult to accept that x can be neglected in comparison with 0.1. If you do feel this, try solving the equation

$$\frac{x^2}{0.1 - x} = 10^{-7}$$

without making this approximation. If you manage to solve this equation correctly, your answer should be 0.9995 × 10^{-4}. Do you think that the approximation is justified?

17. The weak acid can be given the same formula as in the answer to question 16. If $[H^+(aq)]_e$ is again taken as x, the concentration of A^- will, this time, be $(0.02 + x)$ while the concentration of HA will be $(0.2 - x)$.

Putting these values into the equilibrium constant expression and ignoring x in comparison with 0.02 and 0.2 gives

$$0.02x/0.2 = 10^{-4.8}; \quad x = 10^{-3.8}$$

and the pH will be 3.8. The key is **B**.

18. The mixing of the solutions will result in the dilution of each to 0.5 M. Putting the known values into the equilibrium constant expression gives

$$\frac{0.5 \times [OH^-(aq)]}{0.5} = 10^{-5}; \quad [OH^-(aq)] = 10^{-5}$$

Using the given value of K_w

$$[H^+(aq)]_e \times 10^{-5} = 10^{-14}$$
$$[H^+(aq)]_e = 10^{-9}$$

The pH will be 9, key **D**.

19. The concentrations of $C_6H_5NH_2(aq)$ and $H^+(aq)$ will both be $x\alpha$ mol dm^{-3}. The final equilibrium concentration of $C_6H_5NH_3^+(aq)$ will be $(x - x\alpha)$ mol dm^{-3}. Putting these values into the equilibrium constant expression gives

$$K_a = \frac{(x\alpha)^2}{x - x\alpha} = \frac{x^2\alpha^2}{x(1 - \alpha)} = \frac{x\alpha^2}{1 - \alpha}$$

which is key **A**.

20. The concentrations of reactants and products would be expected to approach constant values which would be reached after time t. This is only the case in **B**.

What were the concentrations of X and Y at the start and what is the approximate value of the equilibrium constant?

21. The equilibrium constant expression for this reaction is

$$K_p = \frac{P_e^2 (NH_3)}{P_e(N_2) \times P_e^3 (H_2)}$$

The addition of the stated masses of extra nitrogen and hydrogen will double the pressure of each of these gases in turn. Both these changes will increase the yield of ammonia as the ammonia pressure at equilibrium will be forced to increase to maintain the original value of K_p. The influence of doubling the hydrogen pressure will have the greater influence on the ammonia pressure as the hydrogen pressure appears to the power 3 in the equilibrium constant expression.

The reaction is exothermic on going from

left to right. This shows that K_p will increase as temperature drops from 450° C to 350°C. This increase in K_p will favour the formation of ammonia at equilibrium. In key C, the influence of doubling the hydrogen pressure and lowering the temperature are both operating to increase the yield of ammonia to the greatest extent. This is the correct key.

22. The equilibrium constant expression for this reaction is the reciprocal of that given in the answer to question 21. Inserting the given values for the partial pressures into the expression gives

$$K_p = \frac{100 \times 10^3}{50^2} = 40$$

The correct key is **E**.

23. The numbers of moles of ethanoic acid and alcohol formed will be equal. Let their equilibrium value be x. The numbers of moles of ester and water remaining at equilibrium will both be $1 - x$. If the volume of the equilibrium mixture is V dm^3, the value of K_c will be related to the equilibrium concentrations by the equation

$$K_c = \frac{\left(\frac{x}{V}\right)\left(\frac{x}{V}\right)}{\left(\frac{1-x}{V}\right)\left(\frac{1-x}{V}\right)} = \frac{1}{4}$$

$$\left(\frac{x^2}{(1-x)^2}\right) = \frac{1}{4}$$

Taking square roots of both sides of this equation gives $x/1 - x = \pm\frac{1}{2}$. The two values of x which satisfy this equation are $\frac{1}{3}$ and -1, the second being a solution that can be disregarded in this present problem. The correct key is **B**.

24. The solubility product expression for calcium carbonate is

$$K_{sp} = [Ca^{2+}(aq)]_e \; [CO_3^{2-}(aq)]_e$$

A pair of values for the equilibrium concentrations of calcium and carbonate ions is given in the question. Substituting these in the solubility product expression gives

$$K_{sp} = 0.2 \text{ mol dm}^{-3} \times 2.5 \times 10^{-8} \text{ mol dm}^{-3}$$

$$= 0.5 \times 10^{-8} \text{ mol}^2 \text{ dm}^{-6}$$

This is the same as the solubility product given in key **C**.

25. The equilibrium constant expression for the dissociation of ethanoic acid is similar to that given in the answer to question 16. The hydrogen and ethanoate ion concentrations at equilibrium can be given the same value, call it x. The ethanoic acid is so little ionized that its concentration will still be very close to the original value of 0.2 mol dm^{-3}. Substituting these values in the expression gives

$$2 \times 10^{-5} \text{ mol dm}^{-3} = x^2/0.2 \text{ mol dm}^{-3}$$

$$x^2 = 4 \times 10^{-6} \text{ mol}^2 \text{ dm}^{-6}$$

$$x = 2 \times 10^{-3} \text{ mol dm}^{-3}$$

The correct key is **B**.

26.
$$K_a = \frac{[H^+(aq)]_e [\text{propanoate}^-(aq)]_e}{[\text{propanoic acid}(aq)]_e}$$

$$= 1.3 \times 10^{-5} \text{ mol dm}^{-3}$$

The propanoic acid and propanoate ion concentrations in the mixture will both be 0.05 mol dm^{-3} as each dilutes the other. The only unknown in the above equation is the hydrogen ion concentration, which must be 1.3×10^{-5} mol dm^{-3}. This corresponds to a pH slightly less than 5 so the correct key must be **D**.

27. The equilibrium constant expression used in the answer to question 26 can be rearranged to

$$[H^+(aq)]_e = 1.3 \times 10^{-5} \frac{[\text{propanoic acid}(aq)]_e}{[\text{propanoate}^-(aq)]_e}$$

This is an example of the form of the equation given in key **D**.

28. The pH at the start is 1 which would be the case if the acid was 0·1 M HCl, key **C**.

29. H$^+$ and OH$^-$ ions react in the ratio 1 to 1. 20·5 cm^3 of base must contain the same number of moles of OH$^-$ as there are moles of H$^+$ in 25 cm^3 of the x M solution.
   The base will be stronger in the ratio 25/20·5 so its molarity will be $(25/20.5)x$ which is a little lower than the $1.25x$ in **E** which would come from $(25/20)x$. There is no need to do the division as the key must be **D**.

30. The change in pH during and after the end point is what would be expected if a weak base is used in this titration. The key must be **A** as all the other bases are fully ionized in aqueous solution.

Test 10

Equilibrium 2 (Redox, $E^⊖$, $\triangle G$, $\triangle H$ and $\triangle S$)

Directions summarized for questions 1 to 11				
A	**B**	**C**	**D**	**E**
1,2,3 only correct	**1,3** only correct	**2,4** only correct	**4** only correct	Some other response or combination of responses is correct

1. The standard electrode potentials of four half-reactions are given below:

$$Sn^{2+} + 2e^- \rightleftharpoons Sn \qquad -0.14 \text{ V}$$
$$Fe^{3+} + e^- \rightleftharpoons Fe^{2+} \qquad +0.77 \text{ V}$$
$$2Hg^{2+} + 2e^- \rightleftharpoons Hg_2{}^{2+} \qquad +0.92 \text{ V}$$
$$\tfrac{1}{2}Br_2 + e^- \rightleftharpoons Br^- \qquad +1.07 \text{ V}$$

Based on this information, which of the following reactions will probably take place?

1 $Sn \quad + 2Hg^{2+} \rightarrow Sn^{2+} \quad + Hg_2{}^{2+}$

2 $2Fe^{2+} \quad + 2Hg^{2+} \rightarrow 2Fe^{3+} \quad + Hg_2{}^{2+}$

3 $2Br^- \quad + Sn^{2+} \rightarrow Br_2 \quad + Sn$

4 $2Fe^{2+} \quad + Br_2 \rightarrow 2Fe^{3+} \quad + 2Br^-$

2. The standard electrode potentials of some changes involving compounds of the element vanadium in acid solution are shown below. Which of these reductions could, theoretically, be brought about by the element zinc for which the standard electrode potential $[Zn^{2+}(aq)|Zn(s)]$ is -0.76 V?

$$E^⊖$$

1 $V(OH)_4^+ + 2H^+ + e^- \rightarrow$
$\qquad\qquad 2H_2O + V(OH)_2^{2+} \qquad +1.0 \text{ V}$

2 $V(OH)_2^{2+} + 2H^+ + e^- \rightarrow$
$\qquad\qquad 2H_2O + V^{3+} \qquad +0.36 \text{ V}$

3 $V^{3+} \qquad + \qquad e^- \rightarrow V^{2+} \qquad -0.25 \text{ V}$

4 $V^{2+} \qquad + \qquad 2e^- \rightarrow V(s) \qquad -1.2 \text{ V}$

3. The e.m.f. of the cell

$$Pt,H_2(g, 1atm)|HCl(aq):CuSO_4(aq)|Cu$$

is independent of the

1 temperature

2 concentration of hydrochloric acid

3 concentration of copper(II) sulphate

4 volume of hydrogen

4. In the Nernst equation

$$E = E^\ominus + \frac{RT}{zF} \log_e \frac{[\text{Oxidized form}]}{[\text{Reduced form}]}$$

which of the following quantities can have both positive and negative values?

1 $E^\ominus$

2 T

3 $\log_e \dfrac{[\text{Oxidized form}]}{[\text{Reduced form}]}$

4 z

5.

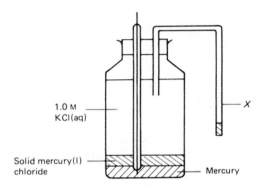

1.0 M KCl(aq)

Solid mercury(I) chloride

X

Mercury

The diagram shows one convenient form of the standard calomel electrode which is used as a reference electrode.

It consists of mercury in contact with solid mercury(I) chloride (Hg_2Cl_2, calomel) and a 1·0 M solution of potassium chloride saturated with mercury(I) chloride.

$$Pt[H_2(g)] \mid 2H^+(aq) \vdots KCl(aq), Hg_2Cl_2(s) \mid Hg$$
$$E^\ominus = +0.28 \text{ V}$$

Which of the following statements is/are correct?

1 To be used as a standard reference electrode, a definite mass of mercury must be present.

2 If more solid mercury(I) chloride is added to the solution, the potential of this half-cell will change.

3 If $E^\ominus$ for the Cu(s) | Cu^{2+}(aq) half-cell is +0·34 V, the $E^\ominus$ value for the following cell

$$Hg \mid Hg_2Cl_2(s), KCl(aq) \vdots Cu^{2+}(aq) \mid Cu(s)$$

will be +0·62 V.

4 The calomel electrode is connected through X to the half-cell whose potential is being measured.

6. For the reaction represented by the equation

$$Cu(s) + 2Ag^+(aq) \rightarrow Cu^{2+}(aq) + 2Ag(s)$$

a value of $\Delta G^\ominus$ could be obtained by

1 setting up Cu(s)|Cu^{2+}(aq) and Ag(s)|Ag^+(aq) half cells (using molar solutions) and measuring the e.m.f. of the resulting cell

2 adding powdered copper to a solution of Ag^+ ions and determining the enthalpy change, repeating this experiment at different temperatures

3 setting up Cu(s)|Cu^{2+}(aq) and Ag(s)|Ag^+(aq) half-cells and determing the molarities of the ions which give an e.m.f. of zero for the cell

4 setting up Cu(s)|Cu^{2+}(aq) and Ag(s)|Ag^+(aq) half-cells, using the resultant e.m.f. to drive a motor and measuring the work which it can do

7. It is found that 8 moles of carbon dioxide, 5 moles of tetrafluoromethane and 3 moles of carbonyl fluoride exist together at equilibrium according to the equation:

$$2COF_2(g) \rightleftharpoons CO_2(g) + CF_4(g)$$

The forward reaction is endothermic.

Which of the following statements are correct?

1 $K_p = K_c = \dfrac{40}{9}$

2 $\Delta G^{\ominus}$ for the forward reaction is negative.

3 The equilibrium position is not affected by pressure changes.

4 The equilibrium position is not affected by temperature changes.

8. The value of the standard free energy change ($\Delta G^{\ominus}$) of a redox reaction at a particular temperature may be calculated solely from measurements of the

1 equilibrium constant of the reaction

2 variation of the rate constant with temperature

3 e.m.f. of the appropriate cell under standard conditions

4 standard enthalpy change ($\Delta H^{\ominus}$) of the reaction

9. The graph represents the standard free energies of formation of the first few members of the alkane series of hydrocarbons. The value for ethylene (ethene) is also shown on the graph. Ethylene is a gas which can be stored at room temperature for long periods without polymerization taking place.

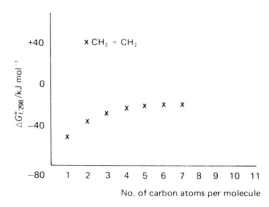

Correct statements about the reaction

$$nCH_2{=}CH_2(g) \rightarrow (-CH_2-CH_2-)_n(s)$$

include that

1 the $\Delta G^{\ominus}_{298}$ value is positive

2 ethene is kinetically stable with respect to its polymers

3 ethene is energetically stable with respect to its polymers

4 the activation energy is likely to have a high value

10. Consider the following reaction:

$$2SO_2(g) + O_2(g) \rightleftharpoons 2SO_3(g);$$
$$\Delta H^{\ominus} = -196 \text{ kJ for the forward reaction.}$$

Which of the following is (are) correct?

1 An increase in temperature from 500 °C to 1000 °C will increase the proportion of sulphur trioxide in the equilibrium mixture.

2 $K_c = \dfrac{[SO_3(g)]^2_{eqm}}{[SO_2(g)]^2_{eqm}[O_2(g)]_{eqm}}$ dm^3 mol^{-1}

3 $K_p = K_c$ for this reaction.

4 The yield of sulphur trioxide can be increased by removing it continuously.

11.

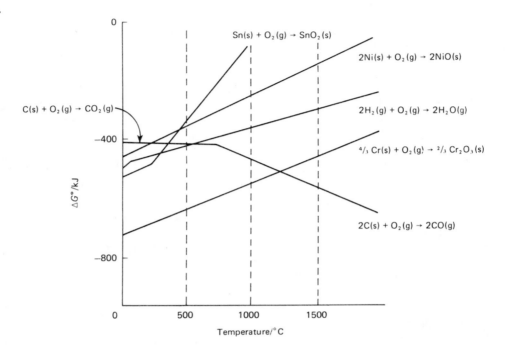

From the Ellingham diagram above, it can be deduced that

1   carbon can reduce chromium(III) oxide at 1000 °C

2   nickel(II) oxide can be reduced by hydrogen at 1500 °C

3   below 300 °C nickel could reduce tin(IV) oxide

4   both hydrogen and carbon could reduce tin(IV) oxide at 500 °C

Directions for questions 12 to 20. Each of the questions or incomplete statements in this section is followed by five suggested answers. Select the best answer in each case.

12. The Nernst equation is

$$E = E^{\ominus} + \frac{0 \cdot 06}{z} \log_{10} [\text{ion}]$$

For the cell

$$\text{Cu(s)}|\text{Cu}^{2+}(\text{aq}) \vdots \text{Pb}^{2+}(\text{aq})|\text{Pb(s)};$$
$$E^{\ominus} = -0 \cdot 47 \text{ V}$$

If the concentration of Pb^{2+} ions is reduced to $0 \cdot 1$ M, the e.m.f. of the cell would change to

A $(-0 \cdot 47 - 0 \cdot 06)$ V

B $(-0 \cdot 47 - 0 \cdot 03)$ V

C $(-0 \cdot 47 + 0 \cdot 03)$ V

D $(-0 \cdot 47 + 0 \cdot 06)$ V

E $(-0 \cdot 47 + 0 \cdot 12)$ V

13. Which of the following reactions, if carried out at constant temperature, is accompanied by a major decrease in entropy?

A $N_2O_4(g) \rightleftharpoons 2NO_2(g)$

B $N_2(g) + 3H_2(g) \rightleftharpoons 2NH_3(g)$

C $CaCO_3(s) \rightleftharpoons CaO(s) + CO_2(g)$

D $2N_2H_4(l) + N_2O_4(l) \rightleftharpoons 3N_2(g) + 4H_2O(l)$

E $H_2(g) + I_2(g) \rightleftharpoons 2HI(g)$

14. For which of the following reactions will the values of $\Delta H_{298}^{\ominus}$ and $\Delta G_{298}^{\ominus}$ be most similar?

A $CCl_4(g) + 2H_2O(g) \rightarrow CO_2(g) + 4HCl(g)$

B $CaO(s) + CO_2(g) \rightarrow CaCO_3(s)$

C $Cu^{2+}(aq) + Zn(s) \rightarrow Zn^{2+}(aq) + Cu(s)$

D $Na(s) + H^+(aq) \rightarrow Na^+(aq) + \frac{1}{2}H_2(g)$

E $C(s) + H_2O(g) \rightarrow CO(g) + H_2(g)$

15. Some standard free energy changes (in $kJ \, mol^{-1}$) for the formation of some ions are:

	$\Delta G_{f,298}^{\ominus}$
$Ag^+(aq)$	77·1
$Fe^{2+}(aq)$	−85·0
$Fe^{3+}(aq)$	−10·7

What is the standard free energy change for the forward reaction

$Ag^+(aq) + Fe^{2+}(aq) \rightleftharpoons Fe^{3+}(aq) + Ag(s)$?

A −151·4

B −18·6

C −2·8

D +2·8

E +18·6

16. For the reaction

$$P + Q \rightleftharpoons R + S$$

$\Delta G_{298}^{\ominus} = -20 \text{ kJ mol}^{-1}$. This tells us that, starting with equimolar quantities of P and Q, in the equilibrium mixture there is

A virtually no P and Q

B P, Q, R and S present with more R and S than P and Q

C P, Q, R and S present in approximately equal amounts

D P, Q, R and S present with more P and Q than R and S

E virtually no R and S

17. For the equilibrium

$$I_2(aq) + I^-(aq) \rightleftharpoons I_3^-(aq)$$

K was found experimentally to be $10^3 \text{ dm}^3 \text{ mol}^{-1}$. Given that R is approximately $8 \times 10^{-3} \text{kJ K}^{-1} \text{mol}^{-1}$ and T approximately 300 K, which of the following is the approximate value for the standard free energy change (in $kJ \, mol^{-1}$) for the reaction?

A +16

B + 8

C 0

D − 8

E −16

18. The following cell is set up:

$$Cu(s) \mid Cu^{2+}(aq) \vdots Ag^+(aq) \mid Ag(s)$$

If the electrode potentials are

$Cu^{2+} \mid Cu$ +0·34 V and $Ag^+ \mid Ag$ +0·80 V

what will be the e.m.f. of the cell?

A +1·14 V

B +0·46 V

C −0·46 V

D −1·14 V

E None of these

19. Standard electrode potentials for the gain of one electron by the ions $Cu^+(aq)$ and $Cu^{2+}(aq)$ are as follows:

$$Cu^+(aq) + e^- \rightarrow Cu(s) \qquad E^\ominus = +0·52 \text{ V}$$
$$Cu^{2+}(aq) + e^- \rightarrow Cu^+(aq) \quad E^\ominus = +0·16 \text{ V}$$

The standard potential, in volts, for the disproportionation

$$2Cu^+(aq) \rightarrow Cu^{2+}(aq) + Cu(s)$$

would be

A −0·68

B −0·36

C +0·36

D +0·68

E +0·88

20.

The diagram in the next column represents a Daniell cell. When the zinc rod and the copper container are connected as part of a completed electrical circuit, a current flows in this circuit. Which of the following statements about this system is NOT correct?

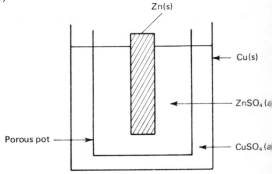

$E^\ominus_{Zn^{2+}\mid Zn}$ = −0·76 V

$E^\ominus_{Cu^{2+}\mid Cu}$ = +0·34 V

A Copper is deposited around the copper container.

B Electrons flow from the zinc to the copper through the external circuit.

C The zinc electrode gradually dissolves.

D If both the solutions are molar, then the cell can produce a maximum e.m.f. of 1·10 volts.

E Sulphate ions migrate through the porous pot from the zinc compartment to the copper compartment.

21. From the following standard electrode potentials

			$E^\ominus/V$
Fe^{3+}	$+ e^-$	$\rightleftharpoons Fe^{2+}$	+ 0.74
I_2	$+ 2e^-$	$\rightleftharpoons 2I^-$	+ 0.55
Sn^{4+}	$+ 2e^-$	$\rightleftharpoons Sn^{2+}$	+ 0.15
Ce^{3+}	$+ 3e^-$	$\rightleftharpoons Ce$	− 2.33

it may be deduced that iodine will

A reduce Fe^{3+} ions to Fe^{2+} ions

B reduce Sn^{4+} ions to Sn^{2+} ions

C reduce Ce^{3+} ions to Ce metal

D oxidize Fe^{2+} ions to Fe^{3+} ions

E oxide Ce metal to Ce^{3+} ions

22. A chart of redox potentials for sulphur and certain of its compounds at pH = 0 is:

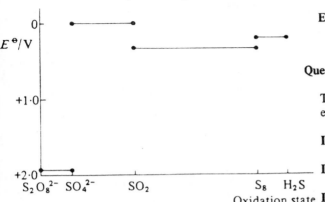

$E^{\ominus}/V$

Which of the following would be expected to undergo a disproportionation reaction at pH = 0?

A $S_2O_8^{2-}$(aq) C SO_2(aq) D S_8(s)

B SO_4^{2-}(aq) E H_2S(aq)

23. Which one of the following would cause the precipitation of more silver in the equilibrium

Ag^+(aq) + Fe^{2+}(aq) ⇌ Ag(s) + Fe^{3+}(aq); $\Delta H = -Q$?

A Warming under atmospheric pressure

B Warming under increased pressure

C Removing some of the solid silver precipitated

D Increasing the concentration of Fe^{3+} ions

E Increasing the concentration of Fe^{2+} ions

24. If the value of ΔG for a reaction is negative, this shows that

A ΔH for the reaction is also negative

B an increase in temperature will favour the formation of products

C the equilibrium constant for the reaction is greater than 1

D the concentrations of reactants will be greater than that of products at equilibrium

E the reaction will take place rapidly at room temperature

Questions 25–27

The standard electrode potentials of some electrodes are as follows:

I Zn^{2+}(aq)|Zn(s)   −0.75 V

II Cd^{2+}(aq)|Cd(s)   −0.40 V

III Ni^{2+}(aq)|Ni(s)   −0.25 V

IV Cu^{2+}(aq)|Cu(s)   +0.32 V

V Ag^+(aq)|Ag(s)   +0.76 V

25. The cell with the smallest e.m.f. is made up of the electrodes

A I and II

B I and III

C I and V

D II and III

E III and IV

26. You have no standard cell available and you calibrate a potentiometer wire 1 metre long so that the balance point for the cell consisting of electrodes I and V is at 75·5 cm. You would then expect the balance point for the cell consisting of electrodes II and III to be at

A 0·15 cm

B 0·65 cm

C 7·5 cm

D 15·0 cm

E 27·0 cm

27. You use filter papers soaked in saturated potassium bromide solution as a bridge between two electrodes. Which electrode would be affected by this bridge?

   A I

   B II

   C III

   D IV

   E V

Questions 28–30

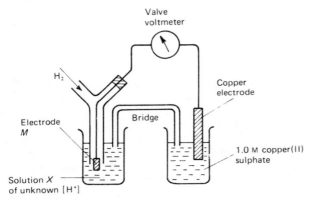

A student set up the above apparatus to determine the hydrogen ion concentration of solution X. The cell is

$$M[H_2(g)] \,|\, 2H^+(aq) \;\vdots\; Cu^{2+}(aq) \,|\, Cu(s)$$

28. The bridge was a glass tube filled with

   A moist cotton wool

   B cotton wool soaked in saturated potassium nitrate solution

   C cotton wool soaked in saturated sugar solution

   D cotton wool soaked in saturated copper(II) sulphate solution

   E cotton wool soaked in alcohol

29. The best material for the electrode M would be

   A polished platinum metal

   B copper metal

   C platinum metal coated with platinum oxide

   D copper metal coated with copper(II) oxide

   E platinum coated with finely divided platinum

30. The e.m.f. for the cell was found to be +0·43 volts. From tables the e.m.f. for the copper electrode was found to be +0·34 volts.

   The e.m.f. of the hydrogen electrode was

   A +0·76 V

   B +0·09 V

   C zero (since the hydrogen electrode is the standard)

   D −0·09 V

   E −0·76 V

Test 10 Answers

1. These half-reactions are already arranged in a suitable form to solve the problem with electrons on the left and with the reactions put in order of decreasing ability to lose electrons. When the arrangement is in this form, the 'anticlockwise rule' can be immediately applied. The reaction in **1** involves the two half-reactions

$$\left(\begin{array}{l} 2e^- + Sn^{2+} \rightleftharpoons Sn \\ 2e^- + 2\,Hg^{2+} \rightleftharpoons Hg_2^{2+} \end{array} \right.$$

In this pair of half-reactions, the higher reaction gives up electrons more readily than the lower and so forces the lower reaction to accept electrons. The higher reaction goes to the left and the lower to the right. When this electron movement occurs, the reaction given by the equation in **1** takes place. **1** is an expected reaction because it involves the anticlockwise movement of electrons — higher to the left, lower to the right.

The reactions in **2** and **4** also involve the anticlockwise movement of electrons in their pairs of half-reactions. The reaction in **3** will not take place in the direction indicated. The reaction predicted by the anticlockwise rule is

$$Sn + Br_2 \rightarrow Sn^{2+} + 2Br^-$$

1, 2 and **4** are correct, key **E**. Why does the question include the word 'probably'?

2. These half-reactions already have electrons on the left but are not arranged in order of decreasing ability to lose electrons. When they are so arranged, the order becomes

$$\begin{array}{l} 4 \\ [2e^- + Zn^{2+} \rightleftharpoons Zn] \\ 3 \\ 2 \\ 1 \end{array}$$

Reaction **4** comes above the zinc reaction and will therefore be capable of reducing zinc ions to zinc. Zinc gives up electrons more readily than

reactions **3, 2** and **1**. The correct key is **A**. Note the use of the word 'theoretically' here, equivalent to the 'probably' of question 1.

$E^{\ominus}$ values can be used to predict the direction of change for electron transfer reactions at 298 K in molar aqueous solution. The rate of reaction cannot be decided, only which direction the reaction will take if the rate is fast enough to be observable. It is, however, possible to predict with certainty whether reaction will *not* take place. If sufficient information is available, it is possible to predict the direction of change at temperatures other than 298 K. Using the Nernst equation (*see* question 4), allowance can be made for solutions that are not 1 molar.

3. The usual $E^{\ominus}$ value for a cell is at 298 K and with 1 molar aqueous solutions. If conditions of temperature and concentration alter, the e.m.f. of the cell, made up of two half-reactions, changes. The alterations in **1, 2** and **3** will change the e.m.f. but, so long as the hydrogen pressure remains at one atmosphere, the volume of hydrogen bubbling through the hydrogen electrode will have no influence on the e.m.f. **4** is the only correct answer to this question and the key is **D**. Which of the changes in **1, 2** and **3** would influence both half-reactions taking place in the cell?

4. The two quantities in the Nernst equation which can have both positive and negative values are those given in **1** and **3**, key **B**.

5. Changes in the quantities of solids do not influence the equilibrium position in a reaction. For this reason **1** and **2** are incorrect.

Look at the cell diagram in **3**. The $E^{\ominus}$ value for the mercury electrode is +0·28 V with respect to the hydrogen electrode. The $E^{\ominus}$ value for the copper electrode is +0·34 V with respect to the hydrogen electrode. This means that the $E^{\ominus}$ value for the copper electrode is +0·06 V with respect to the mercury electrode. The $E^{\ominus}$ value for the cell, following the convention of quoting the polarity of the right-hand electrode with respect to the left, is 0·06 V. **3** is incorrect. Only the statement in **4** is correct and the key is **D**.

6. The cell in **1** will give a value for $E^{\ominus}_{cell}$ from which $\Delta G^{\ominus}$ can be obtained by use of the relation $\Delta G^{\ominus} = -zFE^{\ominus}$. What is the value of z in this case?

The enthalpy change in a reaction is almost independent of temperature over a considerable temperature range. If a phase change occurs for a reactant or product, there is a sudden change in ΔH but otherwise there is just a small trend in value owing to the difference in heat capacities of reactants and products. Experiment **2** will not give a value of $\Delta G^{\ominus}$.

The molarities of $Cu^{2+}(aq)$ and $Ag^{+}(aq)$ which give a cell e.m.f. of zero will be a pair of equilibrium values. These can be put into the experimental equilibrium constant expression:

$$K_c = \frac{[Ag^+(aq)]^2_e}{[Cu^{2+}(aq)]_e}$$

The value of K_c obtained, the experimental equilibrium constant, can be used to find $\Delta G^{\ominus}$ for the reaction using the relation

$$\Delta G^{\ominus} = -2 \cdot 3RT \log_{10} K$$

[The K in the equation above, the thermo-dynamic equilibrium constant, is really equivalent to K_p but for reactions in aqueous solution where no gases are involved there is negligible volume change and K_p and K_c can be taken to be numerically identical]

If the cell is made to produce electrical energy by electron flow, as opposed to having its maximum potential to produce energy measured, the cell p.d. will drop. The work obtained from the cell will be less than the maximum value, $\Delta G^{\ominus}$. The incorrect value will be further affected by energy losses in the motor and in whatever device is used to measure the work done by the motor.

1 and **3** are correct, key **B**.

7. There is the same number of moles of gaseous molecules on both sides of the reaction equation so this is a case where K_p and K_c have the same value. This value will be

$$\frac{8 \times 5}{3^2} = \frac{40}{9}$$

1 is a correct statement.

To decide on the second statement it is necessary to use the relation

$$\Delta G^{\ominus} = -2 \cdot 3RT \log_{10} K_p$$

K_p is greater than unity so $\log_{10} K_p$ is positive. The right hand side of the relation will be negative. $\Delta G^{\ominus}$ must also be negative and **2** is a correct statement.

The third statement is also correct as the number of moles of gas is the same for reactants and products. The fourth statement would only be correct if ΔH for the reaction is zero. This is a most unlikely situation especially as $\Delta G^{\ominus}$ and $\Delta H^{\ominus}$ would be expected to be very similar for this reaction (*see* the answer to question 14). The fourth statement can be taken to be incorrect and the key is **A**.

8. **1** is correct as $\Delta G^{\ominus}$ can be found from the relation $\Delta G^{\ominus} = -2 \cdot 3RT \log_{10} K_p$. Variation of the rate constant with temperature will provide a value for the activation energy of the reaction but not $\Delta G^{\ominus}$. Variation of the rate constant in both directions will give the activation energies of forward and back reactions from which ΔH can be found but not $\Delta G^{\ominus}$. The rate constants for forward and back reactions would, by their ratio, give a value for the equilibrium constant and hence $\Delta G^{\ominus}$ but the second statement does not cover this.

The e.m.f. of the appropriate cell will give the value of $E^{\ominus}_{cell}$ which is related to $\Delta G^{\ominus}$ by the equation of $\Delta G^{\ominus} = -zFE^{\ominus}$ so **3** is correct. The value of $\Delta H^{\ominus}$ on its own would not give a value for $\Delta G^{\ominus}$ – the value of $\Delta S^{\ominus}$ would also be required.

1 and **3** are correct and the key is **B**.

9. From the graph, the limiting value of $\Delta G^{\ominus}_{f, 298}$ for a long-chain hydrocarbon is about -20 kJ mol^{-1}. Putting this value and the corresponding one for ethene into the equation gives

$$nCH_2 = CH_2 (g) \rightarrow (-CH_2 - CH_2 -)_n (s)$$

$\frac{\Delta G^{\ominus}_{f,298}}{\text{kJ mol}^{-1}}$	$n \times +40$	-20

This shows that $\Delta G^{\ominus}_{298}$ for the reaction is -20 $-40n$ which is negative so the first statement is incorrect.

The information given about the long period stability of ethene shows that it is kinetically stable with respect to its polymers. **2** is correct. The large negative value for $\Delta G^{\ominus}_{298}$ shows that, energetically, formation of the polymer is highly favourable and **3** is incorrect. The great kinetic stability of ethene with respect to its polymers shows that the activation energy for polymerization is likely to be high. **4** is correct as is **2** so the key is **C**.

The word 'likely' is used in **4** because one only really knows from the high kinetic stability that the rate constant for polymerization is very low. The rate constant is likely to be low because of high activation energy but it could be low for other reasons. Do you know what one of these other reasons is?

10. This is an exothermic reaction, so raising the temperature will favour decomposition of SO_3. **1** is incorrect.

Everything is satisfactory about the expression for K_c, including the units, and **2** is correct. K_p and K_c are not the same for this reaction as the number of gaseous molecules is different on the two sides of the equation. **3** is incorrect.

If the reaction approaches equilibrium and SO_3 is then removed, the reaction will have to approach equilibrium again, making more SO_3 as it does so. Continuous removal of SO_3 will be equivalent to a constant repetition of this process and the yield of SO_3 will be increased. **4** and **2** are correct, key **C**.

11. The Ellingham diagram shows half-reactions competing for oxygen in the same way that a table of $E^{\ominus}$ values shows half-reactions competing for electrons. The similarity goes further than this. Oxygen is placed on the left of each half-reaction and the more negative $\Delta G^{\ominus}$ values are placed lower in the graph. This makes it possible to apply an anticlockwise rule similar to that used to make predictions using $E^{\ominus}$ values. The Ellingham diagram differs from a table of $E^{\ominus}$ values in that it has a temperature axis. Unless high pressures are used, $E^{\ominus}$ values are very limited in temperature variation by the

small liquid range of water. There is no such limitation on $\Delta G^{\ominus}$ values for competition for oxygen and the temperature in an Ellingham diagram may go up to several thousand degrees. It is in this high-temperature region that information is frequently required about the possibility of removing oxygen from oxides in the extraction of metals.

For each possible deduction in the question, the two half-reactions and their relative positions have been taken from the Ellingham diagram. The direction of the arrows shows the direction of reaction predicted by applying the anticlockwise rule. This is also the direction in which $\Delta G^{\ominus}$ for the reaction is negative.

1 $\quad O_2 + 2C \rightleftharpoons 2CO$
$\quad O_2 + \frac{4}{3}Cr \rightleftharpoons \frac{2}{3}Cr_2O_3$

CO will go to carbon and chromium to Cr_2O_3 so deduction **1** is incorrect

2 $\quad O_2 + 2Ni \rightleftharpoons 2NiO$
$\quad O_2 + 2H_2 \rightleftharpoons 2H_2O$

Nickel oxide will go to nickel and hydrogen to water so deduction **2** is correct.

3 $\quad O_2 + 2Ni \rightleftharpoons 2NiO$
$\quad O_2 + Sn \rightleftharpoons SnO_2$

Nickel oxide will go to nickel and tin to tin(IV) oxide so deduction **3** is incorrect

4 $\quad O_2 + Sn \rightleftharpoons SnO_2$
$\quad O_2 + C \rightleftharpoons CO_2$
$\quad O_2 + 2H_2 \rightleftharpoons 2H_2O$

Tin(IV) oxide will go to tin while carbon and hydrogen will form oxides so deduction **4** is correct

2 and **4** are correct deductions, key **C**.

12. In this problem, z is 2 and $\log_{10} [Pb^{2+}(aq)]$ is -1. The electrode potential of the right-hand electrode will change by

$$\left(\frac{0.06}{2} \times -1 \right) V$$

This is a change of $-0.03V$, making the key **B**.

13. Large changes in entropy occur when there is a change in the number of gaseous molecules as reaction takes place. This arises because only gaseous molecules possess the large number of very closely spaced energy levels of movement (called translation). This gives such a vast number of ways in which gaseous molecules can hold their internal energy that such molecules have high entropy values. The entropy of solids and liquids are both small relative to the entropy of gases. If the number of gaseous molecules decreases during a reaction, there is a major decrease in entropy. This occurs in **B** where four gaseous molecules become only two. Can you spot the reaction where the entropy change will be very small?

14. The connection between the two quantities in the question is

$$\Delta G^\ominus_{298} = \Delta H^\ominus_{298} - T\Delta S^\ominus_{298}$$

(*T* is 298 in this equation)

$\Delta S^\ominus_{298}$ represents the entropy change in the reaction. The ΔG and ΔH values will be very similar if the ΔS value is small.

   A, B, D and **E** involve a change in the number of gaseous molecules so the entropy change will be large. The correct key is **C**, in which gaseous molecules are not involved at all.

15. This problem can be solved in the same way as for a problem where ΔH values are involved rather than ΔG values (*see* the answer to test 5, question 10). Put the $\Delta G^\ominus_{f,298}$ values (in kJ mol^{-1}) below the chemical equation and find whether the total for the products is more positive or more negative than for the reactants. Note that the standard free energies of formation for elements are taken as zero.

$$Ag^+(aq) \;+\; Fe^{2+}(aq) \;\rightleftharpoons\; Fe^{3+}(aq) \;+\; Ag(s)$$

$$\underbrace{+77{\cdot}1 \qquad\quad -85{\cdot}0}_{-7{\cdot}9} \quad\xrightarrow{\hspace{2cm}}\quad \underbrace{-10{\cdot}7 \qquad\quad 0}_{-10{\cdot}7}$$

$\Delta G^\ominus_{298}$ is $-2{\cdot}8$ kJ mol^{-1}, key **C**.

16. A value for $\Delta G^\ominus_{298}$ of -20 kJ mol^{-1} indicates that the equilibrium constant is considerably greater than unity. Reaction will go almost to completion and the key is **A**.

   This arises because

$$\Delta G = -2{\cdot}3RT \log_{10}K_p$$

which gives

$$\log_{10}K_p = -\frac{\Delta G}{2.3RT}$$

$$= \frac{20\,000 \text{ J mol}^{-1}}{2{\cdot}3 \times 8{\cdot}3 \text{ J mol}^{-1}\text{ K}^{-1} \times 298 \text{ K}}$$

The value of $\log_{10}K_p$ works out at almost 4 which means that K_p is approaching 10 000.

17. This question is testing whether you know and can use the equation linking the quantities given.

$$\Delta G^\ominus_T = -2{\cdot}3RT \log_{10}K_p$$

$$\Delta G^\ominus_{300} = -2{\cdot}3 \times 8 \times 10^{-3} \text{ kJ mol}^{-1}\text{ K}^{-1}$$
$$\times 300 \text{ K} \times 3$$

$$= -16 \text{ kJ mol}^{-1} \text{ approx. (key E)}$$

Only approximate values are being used to calculate an approximate ΔG value. For this reason it does not matter that the equilibrium constant given in the question is a K_c rather than a K_p value, especially as the volume change in the reaction is almost zero.

18. The difference between left and right hand electrodes is $0{\cdot}46$ V. The convention is to quote the polarity of the right hand electrode which, in this case, is positive with respect to the left. The key is **B**.

19. The cell diagram for a suitable cell in which to carry out the disproportionation reaction is given below. It is shown this way round, rather than the reverse way, because the reaction in the problem is the reaction through the cell from left to right. Reading the diagram from left to right, it says 'copper(I) goes to copper(II) and copper(I) goes to copper'.

$$Pt \mid Cu^+(aq), Cu^{2+}(aq) \vdots Cu^+(aq) \mid Cu(s)$$

+0.16 V

with respect to hydrogen electrode

+0.52 V

with respect to hydrogen electrode

The right hand electrode is more positive than the left by 0.36 V. By convention, the polarity of the right hand electrode is quoted so the standard potential is +0.36 V, which is key **C**.

20. The first four statements are correct. The final statement must be wrong because migration of ions through the porous pot is meant to compensate for the flow of negative electrons from zinc to copper in the external circuit. If negative sulphate ions are migrating, they must do so into the zinc compartment. The key is **E**.

21. It would be helpful to rearrange the four reactions in order of decreasing ability to lose electrons. This requires complete reversal of the table.

$$E^{\ominus}/V$$

$$Ce^{3+} + 3e^- \rightleftharpoons Ce \quad -2.33$$
$$Sn^{4+} + 2e^- \rightleftharpoons Sn^{2+} \quad +0.15$$
$$I_2 + 2e^- \rightleftharpoons 2I^- \quad +0.55$$
$$Fe^{3+} + e^- \rightleftharpoons Fe^{2+} \quad +0.74$$

The anti-clockwise rule can now be applied. It shows that as iodine goes to iodide ions it can oxidize Ce to Ce^{3+} and Sn^{2+} to Sn^{4+}. The first of these changes is that stated in key **E**.

22. Disproportionation takes place when an intermediate oxidation state is converted, by self oxidation and reduction, to a higher and a lower oxidation state. If you are provided with an $E^{\ominus}$/oxidation state chart, you need to look for a pair of values that are related in the following way:

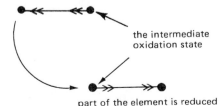

part of the element is oxidized

the intermediate oxidation state

part of the element is reduced

A pair of values with this arrangement is found in the question for SO_2. The chart suggests that this can oxidize itself to SO_4^{2-} while, at the same time, reducing itself to S_8. The correct key is **C**.

23. The reaction is exothermic on going from left to right. This indicates that the formation of silver is increased if the temperature is reduced. Warming will not favour the formation of silver, nor will increasing the pressure as this change can only influence the equilibrium position in a gaseous reaction. A and B are not correct.

As silver is a solid, it does not appear in the equilibrium constant expression for the reaction. Addition or removal of silver will have no influence on the equilibrium position, provided some silver is present. C is not correct.

Increasing the concentration of Fe^{3+} ions will drive the equilibrium from right to left and so reduce the quantity of silver present. Increasing the concentration of Fe^{2+} ions will, however, push the equilibrium in the required direction. The correct key is **E**.

24. The sign of ΔG does not indicate the sign of H with certainty. The two quantities are connected by the relationship

$$\Delta G = \Delta H - T\Delta S$$

The unknown value of the $T\Delta S$ term means that key **A** may not be correct. It is more likely to be correct the more negative the value of ΔG. To predict the influence of temperature on equilibrium position, the sign of ΔH needs to be known so key B cannot be correct. Can you see another reason why the statement made in B is likely to be incorrect?

ΔG is an important quantity because it indicates the value of the equilibrium constant at the temperature for which ΔG is known.

$$\log_{10} K_p = -\frac{\Delta G}{2.3 RT}$$

If ΔG is negative, $\log_{10} K_p$ must be positive and hence K_p will be greater than 1. This is stated in key **C**, the correct key. The statement in D says the opposite. The sign and magnitude of ΔG indicates only the position of final equilibrium, not the rate at which it is attained. Key **E** is also incorrect.

25. **II** and **III** will have the smallest e.m.f. which will be 0·15 V. The key is **D**.

26. The e.m.f. of the cell consisting of **I** and **V** is 1·51 V. There is a voltage drop of 1·51 V along 75·5 cm of wire. This is 0·02 V per cm. The e.m.f. of the cell consisting of **II** and **III** is 0·15 V. This should give a balance point of

$$\frac{0 \cdot 15 \text{ V}}{0 \cdot 02 \text{ V cm}^{-1}} = 7 \cdot 5 \text{ cm}$$

The key is **C**.

27. The potassium ions in the salt bridge will have no effect on any electrode system but the bromide ions will react with silver ions to give a precipitate. The key is **E**.

28. The salt bridge must be composed of conducting ions which are unreactive with the solutions they connect. Aqueous potassium nitrate is a suitable chemical and the cotton wool will slow down mixing of the solutions. The key is **B**.

29. Electrode M is part of the normal hydrogen electrode where the material is that given in **E**.

30. Had the hydrogen electrode been the standard electrode, the e.m.f. of the copper electrode would have been +0·34 V. The difference between copper and the hydrogen electrode has increased by 0·09 V. This must be because the hydrogen electrode has moved further away from the copper value by becoming −0·09 V, key **D**.

Test 11

s- and p-Block elements

Questions 1–5 concern the following compounds:

A lithium bromide

B sodium fluoride

C potassium iodide

D caesium chloride

E barium nitrate

Select, from **A** to **E**, the compound which would behave as described in each question below.

1. Manganese(IV) oxide was mixed with the solid compound and a few drops of concentrated sulphuric acid were added. The gases evolved on warming were passed into silver nitrate solution, giving a white precipitate which dissolved readily in dilute ammonia solution.

2. Dilute sulphuric acid was added to a solution of the compound and a white precipitate was formed.

3. A solution of the compound was treated with a few drops of silver nitrate solution and a yellow precipitate formed which was insoluble in concentrated ammonia solution.

4. When chlorine, dissolved in tetrachloromethane, was shaken with an aqueous solution of the compound, the lower layer darkened; with bromine in tetrachloromethane, no darkening occurred.

5. The solid compound was heated alone in a test tube and a brown gas was evolved; on cooling with a freezing mixture, this gas liquefied.

Questions 6–8 concern the following types of hydrides:

A absorption (i.e., interstitial)

B ionic

C stable, covalent without hydrogen bonds

D stable, covalent with hydrogen bonds

E unstable covalent

Select, from **A** to **E**, the type of hydride which is possessed at room temperature by each set of elements below.

6. Oxygen and fluorine

7. Lithium, potassium and calcium

8. Germanium, tin and lead

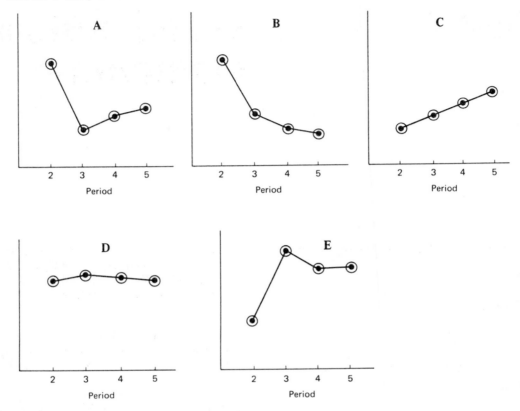

Questions 9–12 concern the above graphs which represent the approximate variation of different properties of elements or their compounds in a Group of the Periodic Table:

Select from **A** to **E** the graph which most closely represents the

9. bond energies of the hydrides of the halogens, fluorine–chlorine–bromine–iodine

10. boiling points of the hydrides of the elements, oxygen–sulphur–selenium–tellurium

11. ionic radius of the Group 2 elements, beryllium–magnesium–calcium–strontium

12. melting points of the dioxides of the Group 4 elements, carbon–silicon–germanium–tin

Directions summarized for questions 13 to 17				
A	**B**	**C**	**D**	**E**
1,2,3	1,2 only	2,3 only	1 only	3 only
correct	correct	correct	correct	correct

13. For the elements Mg, Ca, Sr, Ba, increase in atomic number is accompanied by increase in

1 basicity of hydroxide

2 thermal stability of carbonate

3 first ionization energy

14. For the elements in the period sodium to argon an increase in atomic number is accompanied by

    1 a decrease in atomic radius

    2 an increase in electronegativity

    3 a decrease in boiling point

15. Chemical changes used in the manufacture of sodium hydroxide by the electrolysis of brine using a flowing mercury cathode include

    1 discharge of hydrogen ions at the mercury cathode

    2 reaction of mercury amalgam with water

    3 liberation of chlorine at the anode

16. Electrolytic reduction of the oxide dissolved in a high-temperature melt is used in the extraction of

    1 aluminium

    2 sodium

    3 magnesium

17. Dilute hydrochloric acid or dilute sulphuric acid can be used to distinguish between the salts

    1 Na_2CO_3 and $NaHCO_3$

    2 Na_2SO_3 and Na_2SO_4

    3 $NaNO_2$ and $NaNO_3$

Directions summarized for questions 18 to 26				
A	**B**	**C**	**D**	**E**
1,2,3 only correct	1,3 only correct	2,4 only correct	4 only correct	Some other response or combination of responses is correct

18. Correct statements about the hydrogen halides include that

    1 they are all coloured

    2 the thermal stability decreases with increasing atomic number of the halogen

    3 they all form soluble silver salts

    4 they all donate protons to water

19. *H*, *M* and *Q* are hydroxy-compounds of the elements *X*, *Y* and *Z* respectively. *X*, *Y* and *Z* are in the same period of the Periodic Table.

    H gives an aqueous solution of pH less than 7.
    M reacts with both strong acids and strong alkalis.
    Q gives an aqueous solution which is strongly alkaline.

    Which of the following statements is/are true?

    1 The three elements are metals.

    2 The electronegativities decrease from *X* to *Y* to *Z*.

    3 The atomic radius decreases in the order *X*, *Y* and *Z*.

    4 *X*, *Y* and *Z* could be phosphorus, aluminium and sodium respectively.

20. Which of the following properties are associated with element number 88? (N.B. Element 86 is radon, an inert gas.)

    1 The sulphate is insoluble in water.

    2 The hydroxide is soluble in water.

    3 The carbonate is insoluble in water.

    4 The element in its compounds has only one oxidation state.

21. Which of the following properties of the elements of Group II (beryllium to barium) increase(s) with increasing atomic number?

    1 Stability of carbonate

    2 Solubility of hydroxide

    3 Reactivity with water

    4 First ionization energy

22. From your knowledge of the alkali metals, correct statements about caesium and its compounds would include that

    1 caesium forms a solid, ionic hydride

    2 caesium nitrate decomposes to the nitrite on heating

    3 caesium reacts violently with cold water

    4 caesium carbonate decomposes readily on heating

23. The addition of concentrated sulphuric acid to potassium iodide is NOT suitable for making hydrogen iodide because the

    1 reaction is too slow

    2 hydrogen iodide is contaminated by reduction products of the sulphuric acid

    3 sulphuric acid is too weak an acid to displace hydrogen iodide from its salt

    4 hydrogen iodide is oxidized to iodine

24. Which of the following properties of the elements chlorine, bromine and iodine increase with increasing atomic number?

    1 Ionization energy

    2 Ionic radius

    3 Bond energy of the molecule X_2

    4 Enthalpy of vaporization

25. Which of the following covalent substances react with water to form ions?

    1 Chlorine

    2 Phosphorus trichloride

    3 Silicon tetrachloride

    4 Tetrachloromethane

26. For the alkali metals, which of the following increases with increasing atomic number?

    1 First ionization energy

    2 Electronegativity

    3 Hydration energy of the univalent ion

    4 Atomic radius

Directions for questions 27 to 30. Each of the questions or incomplete statements in this section is followed by five suggested answers. Select the best answer in each case.

27. The highest occupied energy level of the Group 2 element radium is $7s^2$. Which of these statements is likely to be INCORRECT?

   A The element will show an oxidation number +2 in all its compounds.

   B The element will decompose water, liberating hydrogen.

   C The hydroxide of the element will be amphoteric.

   D The carbonate of the element will be very sparingly soluble in water.

   E The carbonate will remain stable up to a higher temperature than calcium carbonate does.

28. Which of the following statements about the elements Mg, Ca, Sr and Ba and their compounds is NOT true?

   A The solubility of the hydroxides in water increases with increasing atomic number.

   B The thermal stability of the carbonates increases with increasing atomic number.

   C The elements all react with water or steam to give hydrogen.

   D The pure chlorides are all liquids at room temperature.

   E The elements invariably form ions of oxidation number +2.

29. Consider the elements carbon to lead in Group 4 of the Periodic Table. As the atomic number increases there is an increase in the

   A stability of the +2 oxidation state

   B acidity of the dioxides

   C first ionization energy of the elements

   D boiling point of the elements

   E stability of the hydrides

30. Which of the following statements is true for hydrogen but for no other element?

   A Hydrogen is never reduced.

   B In reacting with other elements, hydrogen forms both positive and simple negative ions.

   C In reacting with other elements, hydrogen forms both covalent and electrovalent bonds.

   D In reacting with other elements, hydrogen forms an ion containing two electrons only.

   E Hydrogen forms electrovalent bonds with alkali metals.

Test 11 Answers

1. The concentrated sulphuric acid will liberate the gaseous hydrogen halides from **A**, **B**, **C** and **D**:

$$X^- + H_2SO_4 \rightleftharpoons HX + HSO_4^-$$

When X^- is bromide or iodide, the hydrogen halide will be considerably contaminated by the products of redox reactions with the concentrated sulphuric acid (*see* the answer to question 23). A reaction with identical equation occurs if X^- is nitrate. Nitric acid vapour will be liberated from **E**.

If manganese(IV) oxide is present, it will oxidize HCl, HBr and HI to the free halogen:

$$4HX + MnO_2 \rightarrow 2H_2O + MnX_2 + X_2$$

The HF and HNO_3 will be unaffected by the MnO_2. It is almost impossible to oxidize fluoride to fluorine (do you know of one method to achieve this?) and nitrogen in nitric acid is already at its highest possible oxidation state, +5.

The substances passed into the aqueous silver nitrate will be HF and HNO_3 (unchanged by the MnO_2) and Cl_2, Br_2 and I_2. The first two will give no sign of reaction. AgF, rather surprisingly for a silver halide, is soluble in water and so is $AgNO_3$! The halogens will produce precipitates of silver halides by reaction of silver ions with the halide ions which arise from hydrolysis:

$$X_2 + H_2O \rightleftharpoons X^- + OX^- + 2H^+$$

Only in the case of Cl_2 will the precipitate be white and this is the only halogen whose insoluble silver salt is readily soluble in dilute ammonia solution. The Cl_2 would have come from caesium chloride, key **D**.

Do you think that the manganese(IV) oxide is essential in this experiment? What would be observed if it were not used?

2. The sulphates of Group 1, like most Group 1 salts, are soluble in water. It is very difficult to find any sparingly soluble salts of Group 1 metals. Do you know of any?

Group 2 salts are far more likely to be insoluble and the white precipitate in the question is $BaSO_4$ formed in what could be thought of as the sulphate test in reverse. The key is **E**.

3. The reaction is

$$Ag^+ + X^- \rightarrow AgX$$

As the silver salt is insoluble, X^- could only come from **A**, **C** or **D**. The fact that the precipitate is yellow indicates bromide or iodide and the insolubility in concentrated ammonia rules out bromide. The key is **C**.

4. The chlorine is acting as an oxidizing agent:

$$Cl_2 + 2X^- \rightarrow 2Cl^- + X_2$$

The darkening points to bromide or iodide which would form bromine or iodine. Had iodide been present, bromine would also have produced darkening by liberation of iodine:

$$Br_2 + 2I^- \rightarrow 2Br^- + I_2$$

The absence of darkening with bromine indicates that bromide is present which must have come from the lithium bromide, key **A**.

5. The evolution of a brown gas suggests NO_2 coming from a nitrate:

$$Ba(NO_3)_2 \rightarrow BaO + 2NO_2 + \tfrac{1}{2}O_2$$

This is confirmed by the way the gas can be liquefied with a freezing mixture. NO_2/N_2O_4 would behave in just this way. The key is **E**.

It would be possible to obtain a trace of gas from the KI by slight air oxidation:

$$2KI + \tfrac{1}{2}O_2 \rightarrow 2K^+, O^{2-} + I_2$$

What will be the colour of this gas? Will it turn to a liquid on cooling? If you were to carry out this reaction, you would find a solid residue that was mainly unchanged KI. What simple test on the solid would confirm that some of the KI had changed as indicated in the equation?

6. H_2O and HF are stable, covalent molecules with hydrogen bonding. The key is **D**. Does one of these elements form another compound with hydrogen? To which of the types does this other hydride belong?

    There is a third element that could have been included in the question with oxygen and fluorine. What is this third element?

7. In these compounds with reactive metals of Groups 1 and 2, hydrogen behaves as a member of Group 7. Hydrogen forms ionic hydrides in which it is present as the hydride ion, H^-. The compounds are Li^+H^-, K^+H^- and $Ca^{2+}2H^-$ and the key is **B**.

8. The hydrides of Group 4 elements are covalent, without hydrogen bonds. The stability and variety of these compounds decreases on going down the Group. The hydrides of germanium, tin and lead could be described as unstable covalent, key **E**.

9. Fluorine forms strong bonds with other elements and this is one of the main factors responsible for the high reactivity of fluorine. Rather surprisingly, however, fluorine forms relatively weak bonds with itself — this is sometimes attributed to repulsion between the three close lone-pairs of electrons on each fluorine atom. The low bond energy of fluorine contributes to the high reactivity of fluorine.

    The strength of the bond between halogen and hydrogen decreases steadily on going down Group 7 from fluorine to iodine. It becomes easier to break the halogen–hydrogen bond on going down Group 7 and this decrease in bond energy is shown correctly in key **B**.

10. On going UP Groups 5, 6 and 7 the boiling points of the hydrides decrease from period 5 to period 3. This is mainly due to a decrease in van der Waals interaction with decreasing size and decreasing number of electrons. The decrease is shown in **A** and **C**.

    Strong, extensive hydrogen-bonding in

NH_3, H_2O and HF reverses this decrease and gives the hydrides of period 2 relatively high boiling points. The correct key is **A**.

11. Atomic radii and ionic radii both increase on going down a Group. This change is shown in curve **C**.

12. There is a very large increase in both melting point and boiling point on going from CO_2 to SiO_2. This reflects the change in structure from small covalent molecule to covalent giant structure.

    The only curve which shows this large increase is **E** which must be the correct key. If you have a data book available, check that the melting points of GeO_2 and SnO_2 are as shown in **E**.

13. On going down any Group in the Periodic Table, the metallic nature of the element increases. Formation of a basic oxide is a metallic property and this property would be expected to increase on going down Group 2 (**1** is correct).

    The thermal stability of the salts of Group 2 increase on going down the Group and this is true for the carbonates (**2** is correct).

    On going down any Group, the first electron to be removed becomes further away from the nucleus. The effect of reduced attraction due to increasing distance from the nucleus is more important than increased attraction due to increase in nuclear charge. The first ionization energy decreases on going down any Group (**3** is incorrect).

    1 and **2** are correct, key **B**.

14. On going across a period, the element's electrons are entering the same 'shell' while the nuclear charge increases. Experience shows that nuclear charge is the most important factor deciding the atomic radius across a period. The atomic radius decreases at least as far as the halogen and **1** is meant to be correct. There is a problem about the atomic radius of the noble gas where the radius is not measured under conditions comparable with those used for earlier elements. The atomic radius of a noble gas is found from deviations from the gas laws and is called the van der Waals radius. This gives a larger value than other methods because gaseous atoms begin to exert a significant influence on each other when they are still a considerable distance apart.

Electronegativity increases across a period from Group 1 to Group 7. Once again the noble gas is a problem but **2** is meant to be correct.

The trend in boiling point across a period is certainly not a simple decrease. The value starts relatively low for a metal and increases from sodium to magnesium to aluminium as the number of electrons per atom used in metallic bonding increases from one to three. The increase in boiling point continues to Group 4 where silicon uses four electrons per atom in forming a giant covalent structure. This is followed by a very large drop to the low boiling points of the remaining elements which are all composed of small molecules.

1 and **2** are correct, key **B**.

15. **1** is incorrect as sodium is discharged in preference to hydrogen when mercury is used as the electrode. **2** and **3** are correct and the key is **C**.

16. All three elements are extracted by electrolytic reduction of a melt but only in the case of aluminium does the melt contain the oxide of the metal. Only **1** is correct, key **D**.

17. Dilute hydrochloric acid and dilute sulphuric acid are strongly ionized and the hydrogen ions react with the anions of the salts of weak acids. The weak acid is liberated and is either detected as such (H_2S from sulphides) or may be detected by the nature of its decomposition products (S and SO_2 from a thiosulphate).

The reactions included in the question are

$$CO_3^{2-} + 2H_3O^+ \rightleftharpoons 2H_2O + H_2CO_3$$
which decomposes to CO_2 and H_2O

$$HCO_3^- + H_3O^+ \rightleftharpoons H_2O + H_2CO_3$$
which decomposes to CO_2 and H_2O

$$SO_3^{2-} + 2H_3O^+ \rightleftharpoons 2H_2O + H_2SO_3$$
which decomposes to SO_2 and H_2O

$$NO_2^- + H_3O^+ \rightleftharpoons H_2O + HNO_2$$
which decomposes to H_2O, NO and NO_2

A carbonate and a hydrogen carbonate cannot be distinguished by this test because both give carbon dioxide by decomposition of the weak acid, H_2CO_3. A sulphite and a sulphate can be distinguished because only the sulphite gives SO_2. A nitrite and a nitrate can be distinguished because only the nitrite produces oxides of nitrogen. **2** and **3** are correct, key **C**.

18. None of the hydrogen halides is coloured — it is the halogens themselves which possess colour (**1** is incorrect). The compounds become less stable to heat with increasing atomic number (**2** is correct). This thermal stability also shows itself in the preparation of the hydrogen halides from their elements. Hydrogen and fluorine explode when mixed, even in the dark at low temperature. At the other end of the scale, hydrogen and iodine require heat or a catalyst before they will react and reaction only continues until a position of equilibrium has been reached.

HF forms a soluble silver salt, AgF, but the other silver halides are insoluble in water. Use is made of their insolubility in the detection and quantitative estimation of halide ions (**3** is incorrect).

All the hydrogen halides are proton donors (**4** is correct).

$$H_2O + HX \rightleftharpoons H_3O^+ + X^-$$

Do you know how the ability to donate protons varies with the hydrogen halides on going down the Group and can you explain why there is such a large change between HF and HCl?

2 and **4** are correct, key **C**.

The formulae of some of the hydroxides formed by the elements of the period sodium to argon are shown on p. 137. The first formula given is that of the hydroxide where atoms of the element are using all their bonds in a particular oxidation state to link to OH. The possibility of water loss between OH groups results in several of these molecules being met, at least in text-books, in the forms shown beside the simple hydroxide formula.

Basic/alkaline hydroxides	Na^+OH^-
	$Mg^{2+}, 2OH^-$

Amphoteric hydroxide	$Al(OH)_3$

Acidic hydroxides	$Si(OH)_4$	$\begin{array}{c} HO \\ HO \end{array} {>} Si {=} O$
		or H_2SiO_3
	$P(OH)_5$	$\begin{array}{c} HO \\ HO{-}P{=}O \\ HO \end{array}$
		or H_3PO_4
	$S(OH)_6$	$\begin{array}{c} HO \quad O \\ {\diagdown}S{\diagup} \\ HO \quad O \end{array}$
		or H_2SO_4
	$Cl(OH)_7$	$\begin{array}{c} O \\ HO{-}Cl{=}O \\ O \end{array}$
		or $HClO_4$

19. The formation of an aqueous solution with a pH less than 7 shows that H is a soluble non-metal hydroxide. X must be a non-metal such as phosphorus which gives rise to acidic hydroxides such as phosphoric acid, H_3PO_4.

The reaction of M with both acids and alkalis shows that it is an amphoteric hydroxide. Y must be an element near the diagonal separating metals from non-metals. It will be an element such as aluminium forming amphoteric $Al(OH)_3$.

The formation of a strongly alkaline solution shows that Q is the hydroxide of a metal in Groups 1 or 2 (or thallium). Z will be an element such as sodium which forms alkaline NaOH.

The first statement is incorrect. Only Y and Z could be metals. X to Y to Z is non-metal to border-line metal to very metallic metal, so the electronegativities will decrease in that order (**2** is correct).

The atomic radius starts relatively high and decreases across a period, so Z will have the largest atomic radius and X the smallest. The statement in **3** is the reverse of the true situation.

X, Y and Z could be phosphorus, aluminium and sodium, in that order, so **4** is correct.
2 and 4 are correct, key **C**.

20. Element 86 is a noble gas and will have complete s and p outer shells or energy levels. This means that the electronic configuration of element 88 will be in its outer levels, $ns^2 np^6 (n + 1)s^2$. It will be an element of Group 2 (s-block) and will be at the bottom of that Group. Features of this Group are

(1) The sulphate is only really soluble with magnesium, becoming less and less soluble on going down. The first statement is correct.
(2) Hydroxides are one of the few compounds which show the reverse trend in solubility with elements at the bottom of the Group having the most soluble hydroxides. The second statement is correct.
(3) The carbonates of all the elements are insoluble. The third statement is correct. Do any elements have soluble carbonates?
(4) The elements of Group 2 show only the oxidation state +2 in their compounds. The fourth statement is correct.

All four statements are correct and the key is **E**.

21. The first three properties listed in the question increase with increasing atomic number for the Group 2 elements. The key is **A**.

22. All the alkali metals form solid, ionic hydrides (**1** is correct). The tendency of the nitrates to decompose, on heating, into the nitrite and oxygen, rather than oxide, NO_2 and oxygen, becomes stronger on going down the Group. The second statement would be expected to be correct. Reactivity increases on going down the Group and **3** would be expected to be correct. The thermal stability of the carbonates increases on going down the Group with lithium carbonate the only one showing appreciable decomposition.

The other carbonates can be decomposed by strong heating in a glass tube but this is by reaction with the glass rather than thermal instability. The fourth statement is likely to be incorrect and the key is **A**.

23. Concentrated sulphuric acid does form some hydrogen iodide in its rapid, vigorous reaction with potassium iodide (**1** and **3** are incorrect). The reaction is not suitable for making hydrogen iodide because of the strong reducing power of HI and iodide ions in acidic conditions (iodine oxidation state -1). Equally important is the strong oxidizing power of concentrated sulphuric acid (sulphur oxidation state $+6$). Various redox reactions take place, two possible equations being

$$2HI + H_2SO_4 \rightarrow I_2 + SO_2 + 2H_2O$$
$$8HI + H_2SO_4 \rightarrow 4I_2 + H_2S + 4H_2O$$

The statements made in **2** and **4** are correct and the key is **C**.

What concentrated acid can be used to prepare hydrogen iodide by reaction with potassium iodide? Check on the laboratory preparation of hydrogen iodide in a reference book if you do not know the answer. What properties of this acid make it suitable for this reaction?

24. First ionization energies decrease on going down every Group (**1** incorrect) but atomic radii and ionic radii always increase on going down a Group (**2** is correct). The bond energy in halogens decreases from chlorine to iodine, as discussed in the answer to question 9 (**3** incorrect). The boiling points and enthalpies of vaporization increase on going down Group 7 and **4** is correct.

2 and **4** are correct, key **C**.

25. The first three substances are hydrolysed to form ions among the products of reaction. Tetrachloromethane is not hydrolysed except by superheated steam or long boiling with alkali:

$$Cl_2 + H_2O \rightleftharpoons H^+Cl^- + HOCl$$
$$PCl_3 + 3H_2O \rightarrow 3H^+Cl^- + H_3PO_3$$
$$SiCl_4 + 2H_2O \rightarrow 4H^+Cl^- + SiO_2$$

The key is **A**.

Two of the reaction products in the above equations are shown in molecular form when they are, in fact, partially ionized. Can you identify these substances and write equations for their ionization?

26. The only property in the list which increases with increasing atomic number is atomic radius so the key is **D**. The other three properties all decrease with increasing atomic number.

27. This is rather similar to the problem about element 88 (question 20). **A**, **B**, **D** and **E** would be expected to be correct properties of the element radium.

Metallic properties increase on going down a Group and across a period towards the left of the Periodic Table. Radium would be expected to be one of the most metallic of metals. It would not, therefore, be expected to be amphoteric. The correct key is **C**.

28. This is a question about Group 2 as a whole rather than about a specific element in the Group. The chlorides of these elements will be composed of giant structures of ions with very high melting and boiling points. **D** is the key of the statement that is the answer to the question.

29. On going down Group 4, the importance of oxidation state $+2$ increases until, with lead, it is the more stable oxidation state. **A** is the correct key. The other properties listed decrease rather than increase on going down the Group.

30. It is probably best to decide first which of the statements are true for hydrogen.

A is untrue for hydrogen. Hydrogen is reduced when it reacts with reactive metals to form compounds such as Na^+H^-.

B is true for hydrogen which forms both H^+ (or H_3O^+) and H^- ions.

C is true for hydrogen which forms covalent compounds such as CH_4 and electrovalent compounds such as Na^+H^-.

D is true for hydrogen which forms the ion H^- which contains two electrons only.

E is true for hydrogen which forms compounds such as Na^+H^- with alkali metals.

Which of the statements **B**, **C**, **D** and **E** is true for hydrogen but for no other element? It looks as though the answer is **B**. No other element forms both positive and simple negative ions. The word 'simple' is important as there are many metals which, as well as forming positive ions, form negative ions which are either compound ions such as MnO_4^- and CrO_4^{2-} or are complex ions such as $Al(OH)_4^-$, $Fe(CN)_6^{3-}$ and $CuCl_4^{2-}$. Do you know the difference between a compound ion and a complex ion?

It is advisable to check the other keys before finally selecting key **B**. Other elements do form both covalent and electrovalent bonds. Chlorine for example, forms CCl_4 and Na^+Cl^-. **C** is incorrect. Two other elements do form an ion containing two electrons only. These ions are Li^+ and Be^{2+}. **D** is incorrect. Other elements form electrovalent bonds with alkali metals. Na^+Cl^- is an example. **E** is incorrect and **B** must be the key.

This is rather a searching question to finish the test and one put in a most original way.

Test 12

d-Block elements

Questions 1–4 concern the following sections of the Periodic Table:

A Group 1

B Group 3

C transition elements

D Group 5

E Group 7

Select, from **A** to **E**, the section into which you would place the elements described in each question below.

1. These elements have atomic radii which change only slightly with increase of atomic number.

2. These elements occur at the bottom of the troughs in a graph of first ionization energy against atomic number.

3. These elements dissolve in water forming a solution of low pH.

4. These elements and their ions rarely take part in complex formation by either receiving or donating electron pairs.

Questions 5–8 concern the following metals from the first *d*-block series in the Periodic Table:

A scandium (atomic number 21)

B titanium (atomic number 22)

C manganese (atomic number 25)

D iron (atomic number 26)

E copper (atomic number 29)

Select, from **A** to **E**, the metal which

5. has the greatest number of unpaired electrons in its atom

6. forms a colourless ion of oxidation state +4

7. forms the smallest cation of oxidation state +2

8. displays the highest oxidation number

Directions summarized for questions 9 to 14				
A	**B**	**C**	**D**	**E**
1,2,3 correct	1,2 only correct	2,3 only correct	1 only correct	3 only correct

9. Technetium, the element below manganese in the Periodic Table, would be expected to have high values for its

1 melting point

2 boiling point

3 density

10. Properties common to the elements manganese, iron, cobalt, nickel and copper include the ready formation by them all of

  1 coloured ions in aqueous solution

  2 oxides of nitrogen on reaction with concentrated nitric acid

  3 chlorides of formula MCl_2 and MCl_3

11. The oxidation state of copper changes when aqueous copper(II) ions react with

  1 $NaOH(aq)$

  2 $Fe(s)$

  3 $KI(aq)$

12. Isomerism could occur in complex ions of formula ($en = NH_2CH_2CH_2NH_2$)

  1

  2

  3

13. Limestone is present in the blast furnace production of iron in order to

  1 provide a source of CaO

  2 remove some impurities

  3 supply CO_2

14. The conversion of pig iron to steel frequently requires the addition of

  1 oxygen or iron oxide

  2 transition elements

  3 silica

Directions for questions 15 to 21. Each of the questions or incomplete statements in this section is followed by five suggested answers. Select the best answer in each case.

---

15. When aqueous sodium hydroxide is added to an aqueous solution of chromium(III) ions, a green-blue precipitate is first formed which re-dissolves to give a green solution. This green colour could be due to

  A $[Cr(H_2O)_6]^{3+}$

  B $[CrO_4]^{2-}$

  C $[Cr(OH)_4]^-$

  D $[Cr(OH)_3(H_3O)_3]$

  E $[Cr_2O_7]^{2-}$

16. When 0·01 mole of a sample of chromium(III) chloride ($CrCl_3,6H_2O$) is treated in aqueous solution with excess of aqueous silver nitrate, 0·02 mole of silver chloride is precipitated.

The most probable formula of the complex ion in this sample is

  A $[Cr(H_2O)_6]^{3+}$

  B $[CrCl(H_2O)_5]^{2+}$

  C $[CrCl_2(H_2O)_4]^+$

  D $[CrCl(H_2O)_3]^{2+}$

  E $[CrCl_2(H_2O)_2]^+$

17. One process for the manufacture of hydrazine (N_2H_4) involves the action of sodium hypochlorite on excess ammonia at 443 K and 50 atm. The yields calculated on hypochlorite are 70 to 80 per cent.
    If there is as little as 1 part per million of Cu^{2+} ion present, yields drop to about 30 per cent. The most likely reason for this is the ability of Cu^{2+} ions to

    A form complex ions with hydrazine or ammonia

    B oxidize hydrazine

    C catalyse side reactions producing other compounds of nitrogen

    D reduce hypochlorite

    E produce precipitates in the alkaline medium provided by the ammonia solution

18. Titanium has the electronic structure $1s^2 2s^2 2p^6 3s^2 3p^6 3d^2 4s^2$. Which of the following suggested compounds of titanium is UNLIKELY to exist?

    A K_3TiF_6

    B K_2TiF_6

    C $K_2Ti_2O_5$

    D K_2TiO_4

    E $Ti(H_2O)_6Cl_3$

19. The electron configuration of atoms of a certain element is $1s^2 2s^2 2p^6 3s^2 3p^6 3d^3 4s^2$. The maximum oxidation number of the element is

    A +6

    B +5

    C +4

    D +3

    E +2

20. Which of the following is NOT characteristic of the transition elements in the series scandium to zinc?

    A The formation of coloured cations

    B The presence of at least one unpaired electron in a d-orbital of a cation

    C The ability to form complex ions

    D The possession of an oxidation state of +1

    E Catalytic activity of the element or of one of its simple compounds

21. Which of these species will NOT act as a ligand in the formation of complexes?

    A $C_6H_5NH_2$

    B CH_3NH_2

    C NH_4^+

    D NH_3

    E $C_2H_4(NH_2)_2$

22. The following is a graph of first ionization energy against atomic number for successive elements in the Periodic Table. The letters are NOT the symbols for the elements.

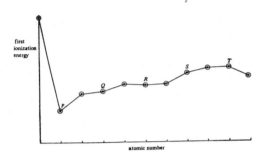

Which element could be manganese (atomic number 25)?

A P

B Q

C R

D S

E T

23. Which of the following molecules is likely to form complex ions with the highest stability constants

A $(C_2H_5)_2O$

B
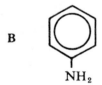
NH$_2$

C $NH_2CH_2CH_2NH_2$

D
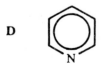
N

E

Questions 24–29 concern the following preparation of potassium hexacyanoferrate(III), $K_3Fe(CN)_6$.

An excess of aqueous potassium cyanide (KCN) is added to a solution of 0·02 moles of iron(II) sulphate.

$$Fe^{2+} + 2CN^- \rightarrow Fe(CN)_2$$

The mixture is boiled in a fume cupboard and the precipitate of iron(II) cyanide then dissolves to form $Fe(CN)_6^{4-}$ ions.

$$Fe(CN)_2 + 4CN^- \rightarrow Fe(CN)_6^{4-}$$

Chlorine is bubbled through this solution and the $Fe(CN)_6^{4-}$ ions are converted to $Fe(CN)_6^{3-}$ ions.

$$Cl_2 + 2Fe(CN)_6^{4-} \rightarrow 2Cl^- + 2Fe(CN)_6^{3-}$$

Crystals of potassium hexacyanoferrate(III), $K_3Fe(CN)_6$, are obtained from this solution and purified by recrystallization.

24. What mass of iron(II) sulphate crystals, $FeSO_4,7H_2O$, should be weighed out to obtain 0·02 moles of iron(II) sulphate? (Fe = 56, S = 32, O = 16, H = 1)

A 1·66 g

B 2·78 g

C 3·04 g

D 3·32 g

E 5·56 g

25. What is the minimum number of moles of potassium cyanide which must be added to the 0·02 moles of iron(II) sulphate to obtain the maximum yield of potassium hexacyano-ferrate(III), $K_3Fe(CN)_6$?

A 0·02

B 0·04

C 0·06

D 0·08

E 0·12

26. The most likely reason for boiling the mixture in a fume cupboard is that

    A potassium cyanide is a covalent compound and would vaporize rapidly on boiling the mixture

    B steam would be prevented from escaping into the laboratory

    C iron(II) sulphate is less likely to be oxidized by atmospheric oxygen to iron(III) sulphate

    D some hydrogen cyanide is formed by hydrolysis and would be given off on boiling the mixture

    E heating of the mixture is more easily controlled by prevention of draughts

27. The oxidation state of iron in $Fe(CN)_6^{4-}$ is

    A −4

    B −2

    C +2

    D +4

    E +6

28. The chlorine is prepared from hydrochloric acid. Before the chlorine is bubbled through the $Fe(CN)_6^{4-}$ solution, any traces of hydrogen chloride must be removed.

    This can best be done by passing the chlorine through wash bottles containing

    A concentrated sulphuric acid

    B water

    C aqueous potassium iodide

    D aqueous sodium hydroxide

    E aqueous chlorine

29. Which of the following is true about the process of recrystallization?

    A Impurities must be insoluble in the solvent chosen.

    B The solid is dissolved in cold solvent and then any insoluble impurities are filtered off.

    C The solution left to crystallize at the end contains no impurities.

    D A solvent is chosen in which the solid to be purified is more soluble at room temperature than at the boiling point of the solvent.

    E The minimum quantity of hot solvent just to dissolve the solid is used.

Questions 30–33

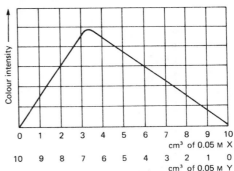

The graph above shows the variation in the colour intensity of different solutions formed by mixing a 0·05 M solution of a metal ion X and a 0·05 M solution of a complexing agent Y, in the proportions indicated on the graph.

30. The most probable formula of the compound formed between X and Y is

    A XY

    B XY_2

    C X_2Y

    D X_3Y_2

    E X_2Y_3

31. If a 0·025 M solution of X had been mixed with the 0·05 M solution of Y and the resulting mixtures analysed as before, then the graph would have been similar to

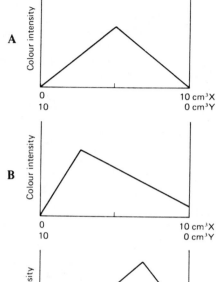

A

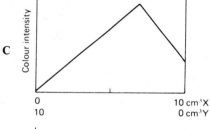

B

C

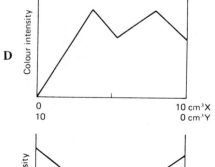

D

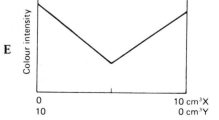

E

32. A metal ion M^{n+} forms a complex ion of formula $[ML_2]^{(n-4)+}$ where L denotes a bidentate ligand. The charge on the ligand L is

A +2

B 0

C −1

D −2

E −4

33. If the bidentate ligand L in $[ML_2]^{(n-4)+}$ is replaced by a neutral monodentate ligand Z, the formula of the resulting complex ion is likely to be

A $[MZ_2]^{2+}$

B $[MZ_2]^{n+}$

C $[MZ_4]^{2+}$

D $[MZ_4]^{n+}$

E $[MZ_6]^{n+}$

Test 12   Answers

1. The elements referred to in this question are the transition elements, key **C**. In the other alternatives between which choice can be made, there is a large increase in atomic radius because of the appearance of a new electron 'shell' with each successive element.

2. The peaks in a graph of first ionization energy against atomic number occur at the noble gases. The graph shows a sudden, large drop at each element following a noble gas. These are the Group 1 elements. The final electron in the atoms of these elements enters a new energy level from which it can be removed relatively easily. There is a pattern of increase and decrease from the Group 1 element to the next noble gas but the value for the first ionization energy never drops below that for the Group 1 element. This is at the bottom of the trough and the key is **A**.

3. The low pH indicates that the aqueous solution is acidic. The only elements which produce acidic solutions are the halogens, key **E**. (Is this really true for iodine?)

$$F_2 \ + \ H_2O \ \rightarrow \ 2H^+F^- \ + \ \tfrac{1}{2}O_2$$
$$Cl_2 \ + \ H_2O \ \rightleftharpoons \ H^+Cl^- \ + \ HOCl$$

4. Did you notice the word 'rarely' in the question? Positive ions of transition elements readily accept lone pairs of electrons in complex formation. Group 5 and Group 7 elements, particularly those of low atomic number, frequently donate lone-pairs from their atoms or ions – NH_3, Cl^- and F^- are ligands. The correct key cannot be C, D or E.

  The high hydration of aluminium salts and the formation of ions such as $Al(OH)_4^-$ indicate that the Al^{3+} ion is able to accept lone-pairs, helped by the strong electrostatic attraction arising from an ion of high charge and small size.

  Key **A** is the best of the choices that are available in the question. It is true that alkali metal ions interact with water molecules when their ionic salts dissolve. Also, some alkali metal compounds are hydrated in the solid state and there is evidence of interaction with ligands such as Cl^- in concentrated aqueous solutions. If you study the electrical conductance of aqueous solutions you will probably have met this interaction. The interactions of this type may be simple electrostatic as opposed to the electrostatic attraction arising from shared electron pairs. Even if electron pairs are involved, these interactions involving alkali metal ions are weaker than any of the others that have to be considered and key **A** is the best choice.

5. The outer electronic configuration of manganese is $...3d^5\,4s^2$ and each of the $3d$ electrons is unpaired. This is the greatest number of unpaired electrons present among the atoms in the list and the key is **C**. Had chromium been in the list, would managanese still have been the correct choice?

6. The transition elements can form colourless ions if the $3d$ energy level is full or empty of electrons. Scandium(III) and copper(I), for example, are colourless. So is titanium(IV), and **B** is the correct key.

7. There is a steady decrease in the ionic radius of the 2+ ion with increasing atomic number in the first d-block series. This can be attributed to the increasing charge and hence increasing attraction of the nucleus for the outer electrons, all of which are in the $3d$ level. **E** is the correct key.

  There is one exception to the trend in the radius of 2+ ions mentioned above. Can you find the exception in a data book?

8. The most significant trend in oxidation states across the first d-block series is the way the maximum value rises by one each element from +3 with scandium to +7 with manganese. +7 is the highest value shown in the whole series and the key is **C**. The maximum value up to this point is equal to the sum of the numbers of $3d$ and $4s$ electrons.

9. A transition element such as technetium would be expected to have high values for all three quantities. The key is **A**. Is there any problem about finding these values for technetium in a data book? What is the lowest value that you can find for a melting point, a boiling point and a density among the transition elements? Is there any significance about where the elements with these low values come in the transition series?

10. All three elements form coloured ions in aqueous solution. They do not all, however, react with nitric acid nor do they all form a chloride MCl_3. The correct key is **D**.

    If you are uncertain about the correctness of **2** and **3**, use a reference book and find a metal in the list that does not react with concentrated nitric acid and also a metal which does not form a chloride MCl_3.

11. Aqueous copper(II) reacts with all three reagents but in the first case a precipitate of copper(II) hydroxide is formed. There is no change in oxidation state here. With iron, copper(II) is reduced to copper metal and with iodide ions, copper(II) is reduced to copper(I). **2** and **3** are correct, key **C**.

12. The first complex ion has no isomers. All six positions around the central chromium are equivalent and it does not matter which of these positions is occupied by the chloride ligand. Optical isomerism is not possible as the ion possesses a plane of symmetry. Can you see this plane of symmetry?

    2 and **3** could occur as optical isomers as both of these ions are completely lacking in symmetry. The key is **C**. Can you sketch the mirror images of **2** and **3**? **3** also has a structural isomer in which the two chloride ligands are further apart. Can you sketch this isomer?

    You are probably familiar with the optical isomerism that arises when a tetrahedral carbon atom is surrounded by four different groups. The complete lack of symmetry in this arrangement makes it possible for optical isomers to occur. **2** and **3** do not contain a tetrahedral carbon atom surrounded by four different groups but optical isomerism is possible because of the lack of symmetry in the structure. The first completely inorganic compound to be

resolved into optical isomers is shown below. The complete lack of symmetry around the central cobalt is very similar to the situation around the chromium in **2**.

13. Limestone breaks down into calcium oxide and carbon dioxide at the high temperature of the blast furnace. The function of the limestone is not to supply carbon dioxide (**3** incorrect) but to provide the basic oxide CaO (**1** correct) which reacts with acidic impurities (**2** correct). The key is **B**.

14. **1** and **2** are correct (key **B**). The oxygen or iron oxide is required to oxidize excess carbon and the addition of transition elements is frequently necessary to give a steel of the required composition. Silica is not added and would need to be removed if present.

15. The precipitate formed when Cr^{3+} ions react with the OH^- ions in aqueous sodium hydroxide is $Cr(OH)_3$. OH^- ions are good donors of lone pairs and these react to dissolve the precipitate as a complex ion, $Cr(OH)_4^-$. The key is **C**:

$$Cr(OH)_3(s) + OH^-(aq) \rightarrow Cr(OH)_4^-(aq)$$

The $Cr(OH)_3$ is behaving here as an electron-pair acceptor (Lewis acid) and the OH^- ion is behaving as an electron-pair donor (Lewis base).

    The complex is frequently given the formula $Cr(OH)_4^-$ but because of the readiness with which Cr^{3+} accepts six lone-pairs (*see* question 12) there are likely to be two water molecules present in the ion. The complex probably has the formula $[Cr(OH)_4(H_2O)_2]^-$.

16. The information shows that one mole of $CrCl_3,6H_2O$ readily provides two moles of chloride ions. It appears that one mole of chloride is strongly held within a complex ion whose formula is likely to be $[CrCl(H_2O)_5]^{2+}$, which is key **B**. The other two chloride ions will be held in an ionic lattice and the remaining water will be held as water of crystallization.

The best of the remaining possibilities is **D**. This will provide two moles of chloride ions per mole of compound and is the only compound, apart from **B**, to be correct on this point. In **D**, however, chromium is surrounded by only four ligands when six would be expected, as in **B**. Three examples of chromium in octahedral complexes are given in question 12. In these complexes, chromium is surrounded by six unidentate ligands in **1**, by three bidentate ligands in **2** and by two unidentate and two bidentate ligands in **3**.

17. One part per million of Cu^{2+} ions causes a drop in yield of 40—50 per cent. This large reduction using such a small concentration of substance indicates that Cu^{2+} is acting as a catalyst for side reactions that do not produce hydrazine. The key is **C**.

18. The outer configuration ...$3d^2 4s^2$ shows that the maximum oxidation state of titanium will be +4. The oxidation state appears to be +6 in K_2TiO_4 and this is the compound unlikely to exist, key **D**.

19. Here the outer configuration is ...$3d^3 4s^2$ indicating that the maximum oxidation state is +5, key **B**.

20. Many transition elements can be obtained, usually with difficulty, in oxidation state +1. The possession of this oxidation state is not, however, a characteristic property of transition elements and the key is **D**.

21. **A**, **B**, **D** and **E** possess one or two lone-pairs of electrons which enable them to act as ligands in forming complex ions. The ammonium ion does not possess a lone-pair and **C** is the correct key. The ammonium ion is a complex ion formed by reaction between the ligand ammonia and the lone-pair acceptor, H^+. Which of the ligands in the problem is bidentate?

22. The large drop in first ionization energy at **P** indicates the appearance of a new electron shell where the final electron to be added is much further from the attraction of the nucleus. P must be an element of Group 1, in this case potassium, atomic number 19. Manganese, atomic number 25, will be six places further on so the correct key is **D**, corresponding to **S**. You could have identified S as manganese if you knew the order of the elements in the Periodic Table after potassium. What elements are *Q, R* and *T*?

23. Stability constants are concerned with the reactions in which water molecules around a central metal ion are replaced by other ligands. High stability constants are associated with powerful ligands which almost completely replace the water molecules. The final molecule in the question does not possess a lone pair and cannot act as a ligand. The other four have lone pairs, 1, 2 diaminoethane, key **C**, being the only molecule with two lone pairs. This allows it to take two 'bites' at the central metal ion and makes it the best choice as the most powerful ligand in the list.

24. The mass of one mole of $FeSO_4,7H_2O$ is 278 g. The mass of 0·02 mole will be $0·02 \times 278$ g which is 5·56 g, key **E**.

25. The equations given suggest that all the cyanide ions used appear in the complex ion $Fe(CN)_6^{3-}$ A minimum of six moles of CN^-, coming from six moles of KCN, will be required for every mole of iron(II) sulphate. 0·02 mole of this salt will require $0·02 \times 6$ mole of KCN. This is 0·12 mole, key **E**.

26. The only sensible suggestion for using the fume cupboard is concerned with the possible evolution of traces of the highly poisonous HCN. This is covered by key **D**.

27. In the complex, each CN is −1 while the charge on the whole ion is −4. Iron must be present as Fe^{2+} which is key **C**.

28. Concentrated sulphuric acid will certainly not remove HCl. The water in the other four suggestions will remove HCl and the presence of

NaOH in the water in **D** will assist very effec-
tively. But the NaOH in **D** and the KI in **C** will
also remove the chlorine by reaction. The only
liquids that will remove HCl while not reacting
excessively with Cl_2 are water (key **B**) and
aqueous chlorine (key **E**).

Aqueous chlorine is slightly better than
water alone because less chlorine will be lost by
dissolving. If the aqueous chlorine were saturated
with chlorine, no more would dissolve and this
would be the ideal liquid to put in the wash
bottles. The key is **E**.

29. **A, B** and **C** are completely untrue statements
about recrystallization. **D** contains an important
word which is very wrong. What is this word? **E**
is a true statement and is the correct key. **E** is
by no means complete but that is not the point
of the question.

30. The greatest concentration of complex, indicated by the highest colour intensity, is close
to the point where 3.4 cm^3 of 0.05 molar X
is mixed with 6.6 cm^3 of 0.05 molar Y. The
mixture contains X and Y in the ratio 1 mole
X to 2 moles Y. The result suggests that, by
moles, twice as much Y as X is required to
form the complex. This would correspond to
a formula XY_2 which is key **B**.

31. The peak colour intensity would be expected
to remain at the ratio 1 mole X to 2 moles Y.
This would be 5 cm^3 of 0.025 molar X to
5 cm^3 of 0.05 molar Y. This, and the shape of
the curve, indicates key **A**.

32. Let the charge on the ligand be x.

The total charge on the complex ion will be
$n + 2x$.

The total charge on the ion is also, from the
given formula, $n - 4$.

$$So \ n + 2x = n - 4$$
$$2x = -4$$
$$x = -2 \ (\text{key } \mathbf{D}).$$

This is equivalent to simply inspecting the
formula and charge on the ions and realising
that two ions of L have made the charge more
negative by four units.

33. The monodentate ligand is neutral so the charge
on the complex must be $n+$. **B, D** or **E** could be
correct.

Two bidentate ligands are likely to be
replaced by four monodentate ligands. This will
give a complex of formula MZ_4^{n+} and the key
is **D**.

Test 13

Carbon chemistry 1

Questions 1–6

The following are classes of organic compounds:

A alkenes

B ketones

C acids

D aldehydes

E esters

Select the class into which you would place a compound which

1. can be reduced to a primary alcohol and forms a silver mirror with Tollens' reagent

2. decolorizes a solution of bromine in tetrachloromethane

3. reacts with water to form an alcohol but has no addition reactions

4. does NOT react with Fehling's solution but forms a crystalline compound with sodium hydrogensulphite (sodium bisulphite)

5. reacts with excess alkaline potassium permanganate solution to form a compound which is further oxidized if the mixture is acidified

6. reacts with HCN to form a compound which is hydrolysed by refluxing with dilute hydrochloric acid to give a compound of formula

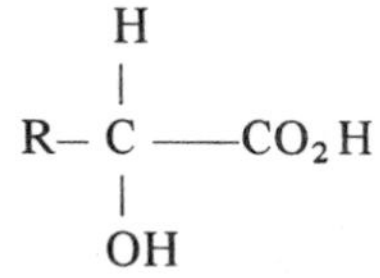

$$R-\overset{\displaystyle H}{\underset{\displaystyle OH}{C}}-CO_2H$$

Questions 7–10 concern the following formulae which represents isomers of the molecular formula $C_4H_{10}O$.

A $C_2H_5OC_2H_5$

B $CH_3CH_2CH(OH)CH_3$

C $CH_3CH_2CH_2CH_2OH$

D $CH_3OCH_2CH_2CH_3$

E $(CH_3)_3COH$

Select, from **A** to **E**, the formula of a compound which

7. on mild oxidation gives a substance which produces a red-brown precipitate on boiling with Fehling's solution

8. reacts with dry phosphorus pentachloride to give 'steamy' fumes, but has no visible effect on warm, acidified potassium dichromate solution

9. on dehydration under suitable conditions gives a mixture of two different alkenes

10. on warming with iodine and aqueous sodium hydroxide forms a pale yellow precipitate

---

Directions summarized for questions 11 to 15

A	B	C	D	E
1,2,3 correct	1,2 only correct	2,3 only correct	1 only correct	3 only correct

---

11. Which of the following could be formed by the dehydration of butan-1-ol, $CH_3-CH_2-CH_2-CH_2-OH$?

    1 But-1-ene: $CH_2=CH-CH_2-CH_3$

    2 But-2-ene: $CH_3-CH=CH-CH_3$

    3 But-1-yne: $H-C\equiv C-CH_2-CH_3$

12. Ammonia would be formed by heating aqueous sodium hydroxide with

    1 $CH_3CH_2NH_2$

    2 $CH_3CH_2NH_3Cl$

    3 CH_3CONH_2

13. A diazonium salt would be formed on adding a solution of sodium nitrite in dilute hydrochloric acid at low temperature to an acidic solution of

    1

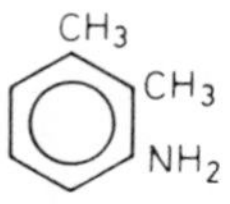

    2

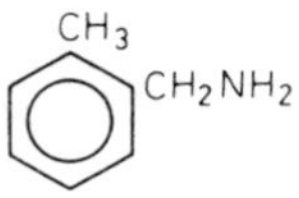

    3

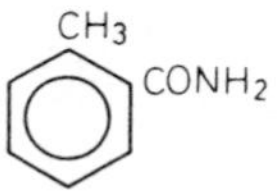

14. Aniline (aminobenzene) is frequently purified by steam distillation. This method is used because aniline is

    1 almost insoluble in water at $100\,^{\circ}C$

    2 able to exert a reasonable vapour pressure at $100\,^{\circ}C$

    3 able to boil below $100\,^{\circ}C$ at atmospheric pressure

15. Hydrogen in the presence of a suitable catalyst can be used to reduce

    1 ethanonitrile (methyl cyanide) to aminoethane (ethylamine)

    2 benzene to cyclohexane

    3 propanoic acid to propanol

Directions for questions 16 to 24. Each of the questions or incomplete statements in this section is followed by five suggested answers. Select the best answer in each case.

---

16. The compound

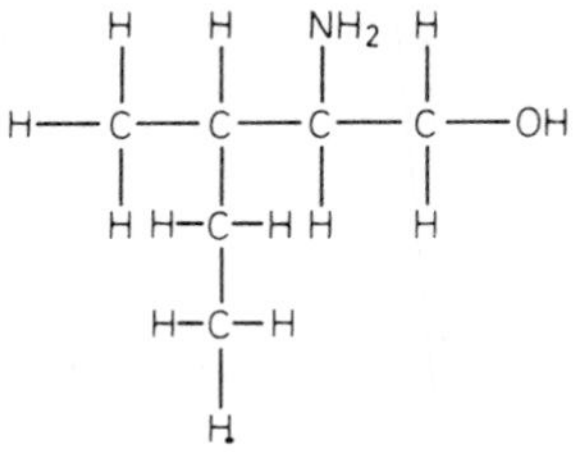

is called

A 2-ethyl-3-aminobutan-4-ol

B 3-amino-2-ethylbutan-4-ol

C 3-methyl-4-aminopentan-5-ol

D 4-amino-3-methylpentan-1-ol

E 2-amino-3-methylpentan-1-ol

17. The experimentally determined melting point of a compound is lower than expected. If each of the following procedures is applied to separate samples of the prepared compound, which would NOT change the melting point?

   A Drying to remove solvent

   B Mixing with pure compound

   C Grinding to a finer powder

   D Recrystallizing from a solvent

   E Adsorbing impurities on charcoal

18. *X*, *Y* and *Z* are three different compounds from the list below. *X* and *Y* react together to form an ester. *X* and *Z* also react to give the same ester as *X* and *Y* but much less readily. *X* reacts with sodium to produce hydrogen and a white solid.

   Compound *Y* could be

   A propanoyl chloride

   B propanoic acid

   C propan-1-ol

   D propanal

   E ethyl cyanide (cyanoethane)

19. Which pair of substances is NOT suitable for the direct preparation, under appropriate conditions, of an ester?

   A Acetic acid and propan-1-ol

   B Acetyl chloride and propan-1-ol

   C Benzoic acid and propan-1-ol

   D Acetyl chloride and phenol

   E Benzoic acid and phenol

20. Which of the following would NOT take place if butan-1-ol were under test?

   A The formation of a yellow derivative on the addition of 2,4-dinitrophenylhydrazine solution

   B The formation of a green colour when warmed with a little acidified potassium dichromate(VI) $(K_2Cr_2O_7)$ solution

   C The production of a sweet smelling compound when heated with a mixture of ethanoic acid (acetic acid) and concentrated sulphuric acid

   D The evolution of hydrogen when sodium is added to it

   E The formation of 1-bromobutane when reacted with sodium bromide and concentrated sulphuric acid

21. 0·15 mole of ethanoic anhydride $[(CH_3CO)_2O]$ was stirred into a small quantity of water until it completely dissolved. The solution was cooled and made up to 1 dm^3 with water. 10 cm^3 of this solution were titrated with 0·1 M sodium hydroxide to an end point with phenolphthalein indicator. What volume of sodium hydroxide, in cm^3, was needed?

   A 3·3

   B 6·7

   C 15·0

   D 30·0

   E 60·0

22. Which one of the following reactions is NOT given by acetone?

    A Reduction of Fehling's solution

    B Iodoform reaction

    C Formation of an addition compound with sodium hydrogensulphite

    D Formation of crystals with 2,4-dinitro-phenylhydrazine

    E Reduction to an alcohol

23. Which of the following compounds would you use in order to obtain a crystalline derivative of an aromatic amine?

    A 2, 4-Dinitrophenylhydrazine

    B Nitrous acid

    C Benzoyl chloride

    D Hydroxylamine

    E Sodium hypochlorite

24. When the colourless liquid chlorobenzene is shaken with bromine water, the chlorobenzene becomes a yellow-orange colour. Which is the best interpretation of this?

    A An addition compound of chlorobenzene and bromine has been formed.

    B The chlorine atom has been replaced by a bromine atom.

    C A hydrogen atom has been replaced by a bromine atom.

    D No multiple bonds are present.

    E The bromine is more soluble in chloro-benzene than in water.

25. Which one of the following is the strongest acid?

    A $CHCl_2CO_2H$ D $CH_2ClCH_2CO_2H$

    B CH_3CO_2H E $CH_3CH_2CO_2H$

    C CH_2ClCO_2H

26. An organic compound can be characterized by the preparation of a crystalline derivative. The derivative *must*

    A have a higher solubility than the original compound in the solvent used

    B have a sharp melting point

    C decompose at its melting point

    D have a low relative molecular mass

    E have a sharp boiling point

27. A colourless liquid undergoes a rapid and exothermic reaction with water giving a strongly acidic solution. The formula of the liquid could be

    A CH_3CH_2COCl D $CH_3CHClCHO$

    B $CH_3CH_2CONH_2$ E $CH_3CH_2CH_2Br$

    C $CH_3CH_2CO_2H$

28. Which of the following reagents can be used to prepare bromoethane from ethanol?

    A Red phosphorus and bromine

    B Concentrated aqueous potassium bromide

    C Dilute hydrobromic acid

    D Bromine and concentrated sulphuric acid

    E Bromine water

Questions 29–32 concern the preparation of an
organic compound which is illustrated below.

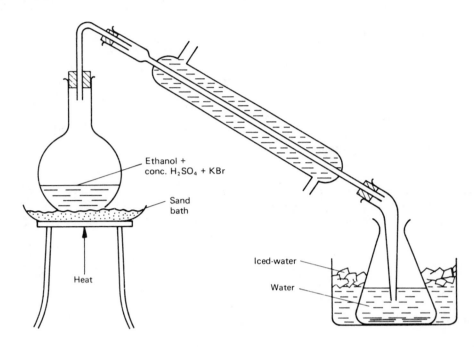

Ethanol +
conc. H_2SO_4 + KBr

Sand
bath

Heat

Iced-water

Water

29. The main organic product of this preparation
    is

    A diethyl ether

    B ethyne (acetylene)

    C ethene (ethylene)

    D ethyl hydrogensulphate

    E bromoethane (ethyl bromide)

30. The reaction mixture in the flask develops a
    light brown colour which becomes black on
    heating. This black colour is due to the

    A formation of ethyl hydrogensulphate

    B presence of carbonaceous impurities in the
       reactants

    C formation of reduction products of the
       sulphuric acid

    D charring of the ethanol by the concentrated
       sulphuric acid

    E production of small quantities of free
       bromine

31. Concentrated sulphuric acid and the inorganic halide react to produce the halogen acid. In addition the sulphuric acid

   A acts as a dehydrating agent

   B increases the rate of the reaction

   C prevents the vaporization of the ethanol

   D maintains the solution at a constant pH

   E reduces the formation of by-products

32. If ethanol and excess concentrated sulphuric acid are heated to about 170 °C in the absence of potassium bromide, the main reaction product would be

   A diethyl ether

   B ethyne (acetylene)

   C ethyl hydrogensulphate

   D ethane

   E ethene (ethylene)

Questions 33–37

   Ethyl ethanoate (ethyl acetate) is a colourless liquid which may be prepared by heating a mixture of ethanol, ethanoic acid (acetic acid) and concentrated sulphuric acid in a flask kept at 140 °C. The ethyl ethanoate distils out of the apparatus.

33. Which apparatus would be most suitable for heating the reaction mixture? (In each case the distillation flask is connected to a condenser and receiver.)

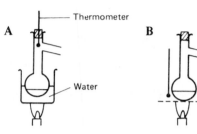

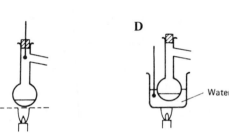

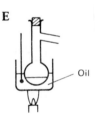

34. The purpose of the concentrated sulphuric acid is to

   A keep the temperature high so that reaction can occur

   B provide H^+ ions which catalyse the reaction

   C oxidize some of the ethanol to ethanoic acid

   D prevent charring of the reaction mixture

   E prevent the ethanol from distilling over

35. The crude ethyl ethanoate which distils over is first shaken with sodium carbonate solution; effervescence can be seen.

The most likely reason for this procedure is to remove

A water which has distilled over as a by-product

B sulphuric acid which has distilled over

C ethanoic acid which has distilled over

D ethanol which has distilled over

E ethene (ethylene) which has been formed by dehydration of ethanol

36. The sodium carbonate solution is then removed from the ethyl ethanoate layer. The ethyl ethanoate is shaken with water which is then removed. The ethyl ethanoate is finally allowed to stand in contact with anhydrous calcium chloride for some time.

The purpose of the anhydrous calcium chloride is to remove

A traces of sodium carbonate which remain

B traces of water before distillation of the ester

C any remaining ethanoic acid from the ester

D any unchanged ethanol

E any unchanged sulphuric acid

37. A sample of the ethyl ethanoate was boiled for some time with excess sodium hydroxide solution in a flask fitted with a vertical (reflux) condenser. When the reaction was complete, the flask was fitted with a condenser in the normal distillation position and a small portion of the product of this reaction was distilled. The distillate was found to be a liquid which reacted with sodium, giving hydrogen. The first portion of the distillate is most likely to contain

A ethyl ethanoate

B ethanoic acid

C sodium ethanoate

D sodium hydroxide

E ethanol

Test 13 Answers

1. The group

H
|
—C=O

present in aldehydes (key **D**) is capable of reduction to a primary alcohol,

H
|
—C—OH
|
H

and oxidation in a silver mirror test to a carboxylic acid,

—C⟨OH / O

2. The group

| |
—C=C—

in alkenes (key **A**) decolorizes bromine dissolved in an organic solvent by addition, forming a dibromoalkane

| |
—C—C—
| |
Br Br

3. Lack of addition reactions rules out **A**, **B** and **D**. An ester (key **E**) is slowly hydrolysed by water to form an alcohol and a carboxylic acid.

4. Aldehydes and ketones can form crystalline compounds by addition reactions with sodium hydrogensulphite. Lack of oxidation by Fehling's solution rules out aldehydes so the key is **B**.

5. The group referred to is the double bond in alkenes, key **A**, which is oxidized by alkaline permanganate to a diol,

| |
—C—C—
| |
OH OH

which would usually be capable of being oxidized further. To what would the diol be oxidized if the alkene is ethene? Can you find a diol that would be difficult to oxidize?

6. Aldehydes and ketones react with HCN by addition to the C=O double bond. In this problem the addition is followed by hydrolysis of the CN group to CO_2H. The R and H of the original aldehyde show up in the final structure and the key is **D**.

7. A primary alcohol (key **C**) forms an aldehyde on mild oxidation and this is capable of being oxidized further to a carboxylic acid in Fehling's test.

8. The 'steamy' fumes indicate HCl gas formed by PCl_5 reacting with an —OH group (key **B**, **C** and **E**). Lack of oxidation by acidified dichromate indicates that the —OH group is tertiary (key **E**).

9. Butan-2-ol (key **B**) can lose water under suitable conditions to form two different alkenes, but-1-ene and but-2-ene. Which other compounds in the list can form alkenes in this way?

10. The formation of a pale yellow precipitate with alkaline iodine is the iodoform reaction. This is given by compounds containing the group

$$CH_3—C⟨^O_R \quad \text{or} \quad CH_3—C⟨^O_H$$

None of the compounds in the problem contains this group. The iodoform reaction is also given by compounds that alkaline iodine is able to oxidize to the necessary group. Butan-2-ol (key **B**) is such a compound as it is readily oxidized to butanone,

$$CH_3—CH_2—C—CH_3$$
$$\overset{||}{O}$$

11. Dehydration of butan-1-ol would be expected to involve loss of water between the first and second carbon atoms to give but-1-ene. Unless rearrangement takes place at the same time, this will be the only product and the key is **D**.

12. **1** will not react with aqueous sodium hydroxide. **2** will react to release the free amine while

158

3 will be hydrolysed to ammonia and the ethanoate ion. Only **3** is correct and the key is **E**.

13. The formation of a diazonium salt occurs only with a primary aromatic amine. **1** is such a compound and the key is **D**. Material **2** is aromatic but it is not an aromatic amine – it is an aliphatic primary amine. What will be formed when this reacts with the nitrous acid present is the solution of sodium nitrite in dilute hydrochloric acid? Material **3** is an amide. What will be formed when this reacts with nitrous acid?

14. The first two statements are correct reasons for choosing steam distillation to purify aniline. The method is used for compounds that boil well above 100 °C but which possess a significant vapour pressure in the region of 100 °C (**3** is incorrect). **1** and **2** are correct and the key is **B**.

    Is it possible to purify aniline by distillation at a temperature below 100 °C by a method that does not involve steam distillation?

15. Hydrogen is capable of reducing nitriles and aromatic compounds if suitable catalysts are used (Pd for the first, Ni for the second). Only **1** and **2** are correct – key **B**. Carboxylic acids are very difficult to reduce. Do you know of one substance that will reduce carboxylic acids?

16. The longest carbon chain including the –OH group contains five carbon atoms so the compound is a pentan-ol. The numbering of the carbon chain should give as low a number as possible to the carbon atom attached to the –OH group. This makes **E** the correct key rather than **C**.

17. The melting point of the compound is lower than expected because impurities are present. **A**, **B**, **D** and **E** will alter the concentration of impurities or remove them completely. This will change the melting point. Grinding to a finer powder will not alter the concentrations of impurities and the melting point will be unchanged. The key is **C**.

18. *X* must be an alcohol which reacts with an acyl chloride (*Y*) and, less readily, with an acid (*Z*) to form an ester. The correct key is **A**. What is the white solid that *X* forms with sodium?

19. Benzoic acid and phenol is the correct pair (key **E**). Phenols do not form esters as readily as alcohols. To what should the benzoic acid be converted so that it will produce an ester by reaction with phenol?

    An acyl chloride reacts rapidly and completely with alcohols to form esters, key **B**. Acyl chlorides also react with phenols to form esters, key **D**. This reaction, which also goes to completion but rather more slowly, is carried out in alkaline solution. Aliphatic and aromatic acids react with alcohols to form esters, keys **A** and **C**. These reactions are slow and reversible. What conditions would you choose for reaction **A** to increase the rate of reaction and improve the yield of ester at equilibrium?

20. **B**, **C**, **D** and **E** are all reactions expected for butan-1-ol. The reaction with 2,4-dinitrophenylhydrazine is given by the carbonyl group in aldehydes and ketones, not by alcohols, and **A** is the correct key.

21. The reaction of ethanoic anhydride with water is hydrolysis. This produces two moles of ethanoic acid from every mole of anhydride. The solution of ethanoic acid formed in the reaction will be 0·3 M. 10 cm^3 of this solution will require 30 cm^3 of 0·1 M NaOH for neutralization (key **D**).

22. Acetone (propanone), like other ketones, is resistant to oxidation and does not give Fehling's reaction. The key is **A**.

23. Benzoyl chloride, key **C**, is the compound that could be used to make a crystalline derivative of an aromatic amine.

24. Chlorobenzene does not react with the bromine in bromine water. It is immiscible with water and, being a better solvent for bromine than is water, will extract the bromine into its own layer. The key is **E**.

25. The acidity of the —CO₂H group can be increased considerably by the presence of chlorine atoms on the carbon atom next to the fuctional group. The molecule in **A** contains two such chlorine atoms and this is the correct key.

26. The value of the sharp melting point of crystalline derivatives can be used to help identify organic compounds. The correct key is **B**. What crystalline derivatives have you made for this purpose?

27. The only molecule to react rapidly and exothermically with water is the acyl chloride, key **A**. What is the acidic product of this reaction?

28. One method of converting alcohols to bromoalkanes is the reaction with red phosphorus and bromine, key **A**. This is equivalent to generating phosphorus tribromide within the reaction mixture, which then reacts with the alcohol. Can you write equations for phosphorus reacting with bromine and for the product reacting with ethanol?

    Questions 29 – 32 are concerned with another way of converting ethanol to bromoethane.

29. The intention of this experiment is to obtain bromoethane, key **E**. **A**, **C** and **D** will be among the unwanted products of the reaction.

30. The darkening described is frequently observed in organic reactions and is caused by the formation of traces of carbon. **B** is incorrect because this suggests that the carbon was present before reaction. **D** is the only suggestion that covers the formation of carbon during the reaction.

31. The reaction of the halogen acid with ethanol produces bromoethane:

$$C_2H_5OH + HBr \rightleftharpoons C_2H_5Br + H_2O$$

This reaction is the reverse of the hydrolysis of an alkyl halide, and concentrated sulphuric acid drives the equilibrium to the right by removing water, key **A**.

32. The initial reaction will form ethyl hydrogensulphate (key **C**) but heating to 170 °C will form either diethyl ether (key **A**) or ethene (key **E**). The correct key is **E** as ethene is formed when the concentrated sulphuric acid is in excess.

33. Note that the details describe 'a flask kept at 140 °C'. This is only possible with the oil bath, key **E**.

34. The concentrated sulphuric acid has two functions — to remove water and to provide hydrogen ions to catalyse the reaction. The second function is described in key **B**.

35. The effervescence with sodium carbonate is due to evolution of CO_2, which suggests that acid is present. The acid must be ethanoic acid (key **C**) rather than sulphuric acid which is much less volatile. Check in a data book the boiling point of ethanoic acid to see whether distillation from a flask kept at 140 °C is likely to make this an important impurity.

36. The purpose of the anhydrous calcium chloride is is to remove water (key **B**) by formation of hydrated calcium ions.

37. Refluxing ethyl ethanoate with aqueous sodium hydroxide will hydrolyse the ester. The products will be ethanol and ethanoate ions. Ethanoic acid (key **B**) would not be driven off on distillation unless the mixture was first acidified. The acid is also less volatile than the other two possibilities which would react with sodium, these being water and ethanol. Water is not included in the choices and ethanol (key **E**) is the correct key. Had both water and ethanol been in the list, which one would you have chosen?

Test 14

Carbon chemistry 2

Questions 1–5 concern the following reactions which are typical of certain functional groups in organic compounds:

A addition of bromine by reaction with a solution of bromine in tetrachloromethane

B formation of hydrogen chloride by the action of phosphorus pentachloride

C evolution of nitrogen by the action of nitrous acid

D formation of tri-iodomethane (CHI_3) by the action of iodine and potassium hydroxide

E formation of an alcohol by boiling with aqueous sodium hydroxide

Select a reaction shown by

1. $CH_3 CH = CHC_2 H_5$

2. $C_6 H_5 CH_2 NH_2$

3. $C_6 H_5 CH_2 OH$

4. $CH_3 CO_2 CH_3$

5. $CH_3 CO_2 H$

Questions 6–9

Organic compounds containing a hydroxyl group can be classified as follows

A primary alcohols, i.e., those containing the $-CH_2 OH$ group

B secondary alcohols $\left(\begin{array}{c} \diagdown \\ \diagup \end{array} CHOH \ \text{group} \right)$

C tertiary alcohols $\left(\begin{array}{c} \diagdown \\ = \\ \diagup \end{array} COH \ \text{group} \right)$

D phenols ($-OH$ group attached direct to a benzene ring)

E carboxylic acids $\left(-C \diagup_{OH}^{O} \ \text{group} \right)$

Select, from **A** to **E**, the category for a compound which

6. has the structural formula

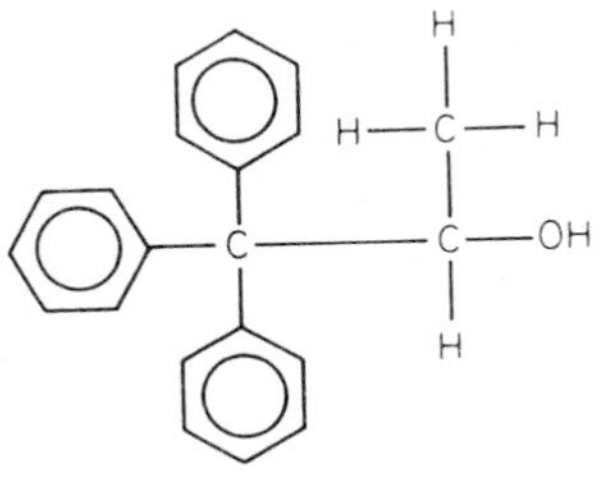

161

7. will react with 3,5-dinitrobenzoyl chloride to give a product of formula

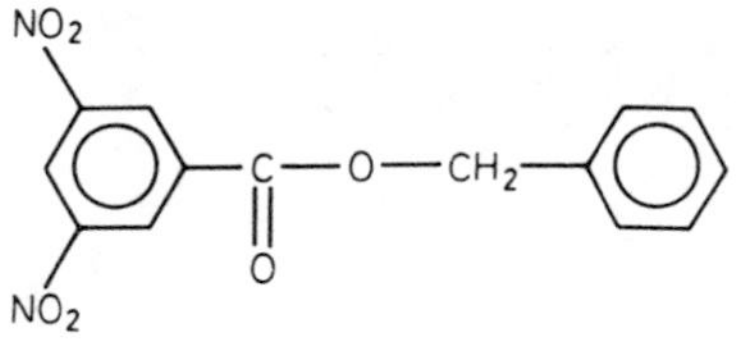

8. is oxidized by acidified dichromate solution to a compound of formula

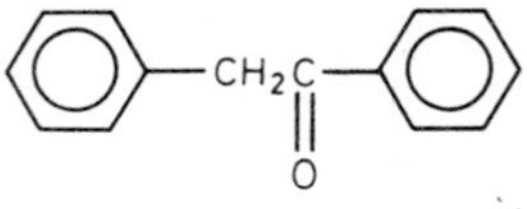

9. CANNOT easily be oxidized and does NOT react with sodium carbonate solution, although its aqueous solution has a pH less than 7

Directions summarized for questions 10 to 16				
A	**B**	**C**	**D**	**E**
1,2,3 only correct	1,3 only correct	2,4 only correct	4 only correct	Some other response or combination of responses is correct

10. An organic compound A of molecular formula $C_4 H_{10} O$ undergoes oxidation to give a compound B of molecular formula $C_4 H_8 O_2$. A could be

1 $CH_3 CH_2 CH_2 CH_2 OH$

2 $CH_3 CH_2 CH(OH)CH_3$

3 $(CH_3)_2 CHCH_2 OH$

4 $(CH_3)_3 COH$

11. On complete oxidation, one mole of an organic compound Y produced three moles of water. Y could be

1 ethanol

2 ethane

3 methanol

4 propene (propylene)

12. Properties desirable in a derivative to be used for identification of a carbon compound include that the derivative should

1 be easily prepared

2 be easily purified

3 be stable below its melting point

4 melt above 100 °C

13. In which pair(s) of reagents will each member of the pair react separately with ethanoic acid $(CH_3 CO_2 H)$?

1 $H_2 SO_4$; NaOH

2 NH_3 ; PCl_3

3 $H_2 SO_4$; PCl_3

4 NH_3 ; Cl_2

14. Which of the following formulae can represent more than one compound?

1 $CH_4 O$

2 $C_2 H_2 Cl_2$

3 $H_2 CO_2$

4 $C_2 H_6 O$

15. Which of the following will undergo acetylation (ethanoylation)?

1 $CH_3CH_2NH_2$

2 CH_3COOH

3 $CH_3CH_2CH_2OH$

4 $\overset{\displaystyle O}{\underset{\displaystyle \|}{CH_3CCH_3}}$

16. Which of the following compounds will form addition products with bromine?

1 Propanone (acetone)

2 Ethane-1,2-diol (ethylene glycol)

3 Aniline (phenylamine)

4 Cyclohexene

Directions for questions 17 to 25. Each of the questions or incomplete statements in this section is followed by five suggested answers. Select the best answer in each case.

17. The compound $(CH_3)_2 C(OH)CH_2CH_3$ is correctly named

A 2-methylbutan-2-ol

B 2-ethylpropan-2-ol

C 4-methylbutan-3-ol

D pentan-3-ol

E 3-methylbutan-3-ol

18. The most suitable method of converting ethanol to iodoethane would be to

A reflux iodine and ethanol

B allow ethanol and potassium iodide to react in the presence of dilute acid

C heat potassium iodide and ethanol with concentrated sulphuric acid

D reflux red phosphorus, ethanol and iodine

E react ethanol with phosphorus pentaiodide in the cold

19. Acetic (ethanoic) anhydride can be most conveniently prepared in the laboratory by

A distilling acetic (ethanoic) acid with phosphorus pentachloride

B passing the vapour of acetic acid over hot calcium oxide

C heating acetic acid with concentrated sulphuric acid

D heating together anhydrous sodium acetate (ethanoate) and acetyl (ethanoyl) chloride

E heating together anhydrous sodium acetate and chloroacetic acid

20. A mixture of 0·1 mole of ethanoic anhydride (acetic anhydride) and 0·1 mole of phenylamine (aniline) was warmed gently until no further reaction occurred and then shaken for some time with water.

The volume of 1·0 M sodium hydroxide solution required to neutralize the aqueous product was

A $0 \ cm^3$

B $25 \ cm^3$

C $50 \ cm^3$

D $100 \ cm^3$

E $200 \ cm^3$

21. One stage in the identification of an aldehyde or ketone is the preparation of the 2,4-dinitro-phenylhydrazine derivative.

Which of the following is NOT a contributing factor in the choice of the 2,4-dinitrophenyl-hydrazone as the derivative?

A The derivatives are brightly coloured and suitable for separation by thin layer chromatography.

B Being crystalline, the derivatives are easily purified by recrystallization.

C The derivatives can be readily identified by determining their melting points, as the melting points are in a suitable temperature range.

D The derivatives do not decompose at or below their melting points.

E The derivatives can be obtained in high yield, so that only small quantities of the carbonyl compounds need be used.

22. Which reagent would react with a $-CH_2OH$ group in an organic compound but NOT with a

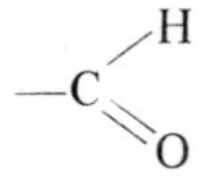

$$-C\overset{\displaystyle H}{\underset{\displaystyle O}{\diagdown}}$$

group to give a simple organic product?

A Acidified potassium dichromate

B Potassium cyanide

C Phosphorus pentachloride

D Sodium metal

E Fehling's solution

23. Propanoamide ($C_2H_5CONH_2$) could be prepared by

A reacting ammonia with methyl propanoate ($C_2H_5CO_2CH_3$)

B adding ethanoyl chloride (CH_3COCl) dropwise, until in excess, to l-aminoethane ($C_2H_5NH_2$)

C bubbling ammonia gas into an ethereal solution of propanal (C_2H_5CHO)

D adding ammonia solution to propan-1-ol ($C_2H_5CH_2OH$)

E reacting ammonia with ethyl ethanoate ($CH_3CO_2C_2H_5$)

24. A certain organic compound reacts vigorously with water to give an acidic solution which instantly yields a precipitate of silver bromide when treated with aqueous silver nitrate.

The compound could be

A CH_2CHCH_2Br
            $\underset{\displaystyle OH}{|}$

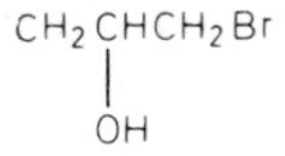

B CH_3CH_2CBr
                    $\underset{\displaystyle O}{\|}$

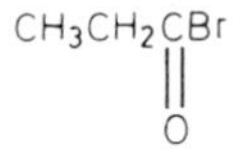

C

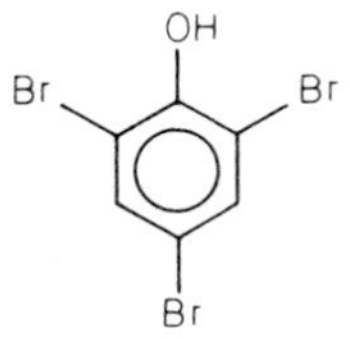

D

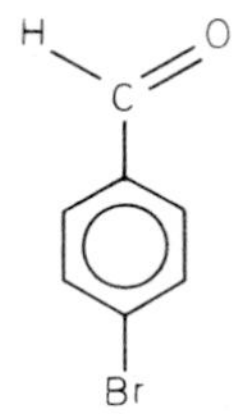

E $BrCH_2CH_2COH$
                      $\underset{\displaystyle O}{\|}$

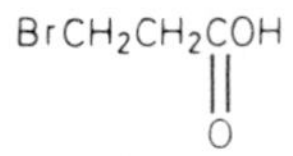

25. A certain organic compound dissolved in water WITHOUT violent reaction and does NOT react readily with sodium carbonate solution.

The compound could be

A 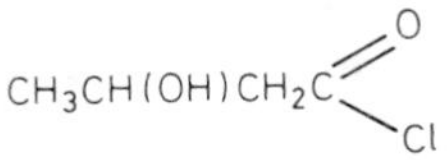

B 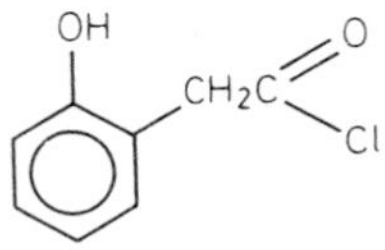

D 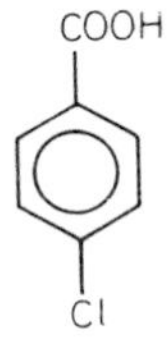

C 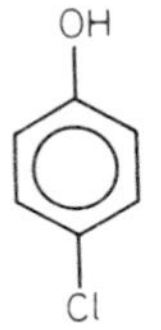

E $CH_3CHClCH_2COOH$

26. An organic compound has the following properties:

I 0.01 mole react with 200 cm^3 of 0.1 M NaOH

II It decolourizes bromine water.

III It loses water readily on heating.

The formula of the organic compound could be

A $CH_2=CHCOCl$

B $CH=CH_2$

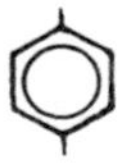

CO_2H

C H — C — CO_2H
      ||
  H — C — CO_2H

D CO_2H
  |
  CO_2H

E HO_2C — C — H
      ||
  H — C — CO_2H

27. A compound X decomposes on heating to yield a volatile product, which forms an oxime.

The formula of X could be

A CH_3CO_2Na **D** $(CH_3CO_2)_2Ca$

B CH_3CONH_2 **E** $CH_3CO_2CH_3$

C HCO_2CH_3

28. Treatment of $CH_3CHBrCH_3$ with solid sodium methoxide (CH_3ONa) would give

A $CH_3CH(ONa)CH_3$ **D** $CH_3CH{=}CH_2$

B $CH_3CH(OH)CH_3$ **E** $CH_3CH(CH_3)_2$

C $CH_3CH(OCH_3)CH_3$

Questions 29–34 concern the preparation of 1,2-dibromoethane, illustrated below.

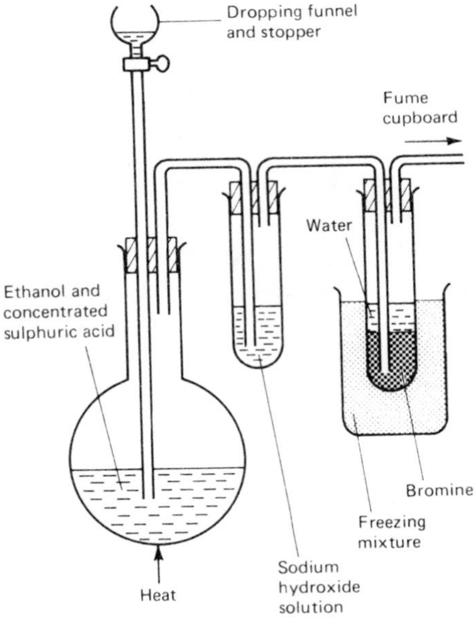

The mixture in the flask was heated to about 170 °C. Ethene (ethylene) was formed and reacted with the bromine in the second test tube to form 1,2-dibromoethane.

$$CH_2{=}CH_2 + Br_2 \rightarrow CH_2BrCH_2Br$$

29. The concentrated sulphuric acid was used in EXCESS in order to

   A prevent the vaporization of the ethanol

   B reduce the formation of diethyl ether

   C increase the rate of the reaction

   D avoid the contents of the flask evaporating to dryness

   E prevent the formation of ethyne (acetylene)

30. The purpose of the sodium hydroxide solution in the central test tube was to absorb

   A any bromine which would otherwise have diffused back into the flask

   B any sulphuric acid which distilled over from the flask

   C any ethanol which distilled over from the flask

   D sulphur dioxide formed by reduction of the sulphuric acid in side reactions

   E 1,2-dibromoethane formed in the reaction

31. The main purpose of surrounding the second test-tube with a freezing mixture was to

   A increase the rate of reaction between the ethene (ethylene) and bromine

   B solidify any ethanol which had distilled from the flask

   C minimize the volatility of the ethene (ethylene)

   D prevent hot ethene (ethylene) from cracking the test-tube

   E minimize the vaporization of the bromine

32. The reaction mixture in the flask developed a light brown colour which became black on heating to about 170 °C. This coloration was due to

   A the formation of ethyl hydrogensulphate

   B the presence of small quantities of diethyl-ether

   C the charring of the ethanol by the concentrated sulphuric acid

   D bromine diffusing back into the flask

   E reduction products of the sulphuric acid

33. The 1,2-dibromoethane is usually contaminated with excess bromine. This could best be removed by washing with

   A cold, dilute aqueous sodium hydroxide

   B hot, concentrated aqueous sodium hydroxide

   C concentrated sulphuric acid

   D distilled water

   E dilute, aqueous sodium chloride

34. The pure 1,2-dibromoethane was tested to show the presence of bromine in the compound. This could best be done by

   A acidifying with dilute nitric acid, adding aqueous silver nitrate and looking for an immediate cream precipitate

   B adding alcoholic silver nitrate and looking for an immediate cream precipitate

   C observing the brown colour of bromine in the pure compound

   D adding chlorine water and looking for the brown colour of bromine

   E warming with aqueous sodium hydroxide, adding acidified silver nitrate solution and looking for an immediate cream precipitate

Questions 35–38 concern the preparation of iodobenzene (C_6H_5I).

I   A mixture of concentrated hydrochloric acid, aniline and an equal volume of water was cooled in an ice bath to about 5 °C. A cooled aqueous solution of sodium nitrite was added slowly keeping the temperature below 10 °C. Benzenediazonium chloride was formed

$$C_6H_5NH_2 + HNO_2 + HCl \rightarrow$$
$$C_6H_5N_2Cl + 2H_2O$$

II   After a few minutes the mixture was transferred to a distillation flask and a solution of potassium iodide added. A vigorous reaction took place. When the reaction subsided, the flask was gently heated until a dark crude oil of iodobenzene separated out. The iodobenzene was separated from the aqueous layer, washed twice with water and once with an aqueous solution of sodium thiosulphate.

$$C_6H_5N_2Cl + KI \rightarrow C_6H_5I + N_2 + KCl$$

III The washed iodobenzene was mixed with water and steam distilled.

IV   The iodobenzene was separated from the aqueous layer and dried.

35. The most probable reason for keeping the temperature below 10 °C in operation **I** was to prevent

A loss of aniline by volatization

B benzenediazonium chloride being formed too quickly

C a reaction between aniline and sodium nitrite

D the formation of nitrous acid

E decomposition of benzenediazonium chloride

36. The most likely reason for washing with sodium thiosulphate would be to remove

A iodine

B sodium nitrite

C potassium iodide

D nitrous acid

E aniline

37. In an actual preparation of iodobenzene, 9·3 g of aniline (relative molecular mass 93) were used and 16·32 g of iodobenzene (relative molecular mass 204) were formed. The percentage yield of iodobenzene is

A   8 per cent

B   50 per cent

C   75 per cent

D   80 per cent

E   100 per cent

38. Diazonium salts are useful in the synthesis of aromatic compounds. Another compound usually prepared from a diazonium salt is

A benzoic acid

B benzylamine

C benzene-azo-2-naphthol

D toluene

E benzyl alcohol

Test 14   Answers

1. The molecule is an alkene and its reactions will involve addition, including the addition of bromine dissolved in an organic solvent, key **A**.

2. This molecule is a primary aliphatic amine (the benzene ring is not attached to the nitrogen atom). Formation of nitrogen by reaction with nitrous acid is a characteristic of such compounds and the key is **C**.

3. This molecule is an alcohol and the reaction with PCl_5 (key **B**) is frequently used to detect an $-OH$ group by the evolution of fumes of HCl.

4. The molecule is an ester and key **E** describes the hydrolysis of such a compound when the products are an alcohol and the sodium salt of a carboxylic acid.

5. This is a carboxylic acid and the presence of an $-OH$ group in such a molecule gives rise to HCl by reaction with PCl_5, key **B**.

6. The structure here is that of a secondary alcohol, key **B**.

7. The molecule reacting with the acid chloride must have the structure

   It is a primary alcohol, key **A**.·

8. The oxidation product is a ketone which will have come from a secondary alcohol, key **B**. What is the structure of this secondary alcohol?

9. This molecule is acidic but not acidic enough to react with sodium carbonate. It must contain the weakly acidic phenol structure, key **D**. Phenols are slightly oxidized fairly readily to give traces of highly coloured compounds but the oxidation is not extensive and it would be reasonable to say, as the question does, 'cannot easily be oxidized'.

10. A is an alcohol that is capable of being oxidized to a carboxylic acid, **B**. A must be a primary alcohol. 1 and 3 are correct, key **B**.

11. The molecule must contain six hydrogen atoms in order to produce three moles of H_2O per mole on complete oxidation. Ethanol, ethane and propene all contain six hydrogen atoms in their molecules and the key is **E**.

12. The first three properties are desirable, key **A**.

13. All the reagents except H_2SO_4 react with ethanoic acid. Both members of the pair in **2** and **4** will react and the key is **C**. One of the reagents requires special conditions before it will react with ethanoic acid. What is this reagent and what are the conditions for reaction?

14. **2** can exist as three isomers and **4** as two isomers. The key is **C**. Can you draw these isomers?

15. Amines such as **1** and alcohols (**3**) will undergo ethanoylation and the key is **B**. This is assuming that ethanoylation means direct introduction of the $-COCH_3$ group by reaction with ethanoyl chloride or ethanoic anhydride. It is possible to introduce the $-COCH_3$ group into ethanoic acid (**2**) by first converting it to its sodium salt and then heating this with ethanoyl chloride.

16. Only cyclohexene (**4**) will react with bromine by addition. The key is **D**. The other three molecules can react with bromine but by substitution. Substitution reactions are frequently slow at room temperature but one of these substances reacts rapidly with bromine by substitution under these conditions. Which substance is this and what is produced?

17. There are four carbon atoms in the longest chain containing the $-OH$ group. The molecule

is a butan-ol. There is a methyl group and an —OH group on the second carbon atom so the key is **A**.

18. Iodine and ethanol (**A**) will not react unless the solution is alkaline and the product then is not iodoethane. What is the product here?

    Ethanol, KI and acid (**B**) would not react unless concentrated sulphuric acid was used (**C**) but then redox reactions between iodine (−1) and the acid would interfere with the reaction that is giving iodoethane:

$$HI + C_2H_5OH \rightleftharpoons C_2H_5I + H_2O$$

D and **E** use the reaction of phosphorus iodides with ethanol, the reaction in **D** involving

$$PI_3 + 3C_2H_5OH \rightarrow 3C_2H_5I + H_3PO_3$$

The instability and ready oxidation of phosphorus iodides make it necessary to generate them within the reaction mixture by the reaction between red phosphorus and iodine. Key **D** is the correct choice.

19. The correct key is **D**, where the reaction is

$$CH_3CO_2Na + ClCOCH_3 \xrightarrow{-NaCl} CH_3CO.O.COCH_3$$

What would be produced in reactions **A** and **B**?

20. The reaction is

The equation shows that the substances react in the ratio 1 mole to 1 mole, forming 1 mole of ethanoic acid. Using the quantities in the question would produce 0·1 mole of ethanoic acid which would require 0·1 mole of sodium hydroxide for neutralization. The sodium hydroxide would be contained in 100 cm^3 of 1·0M solution. The key is **D**.

21. Neither the colour nor the possibility of separation by thin layer chromatography are considered when the choice is made of the 2,4-dinitrophenylhydrazone as a suitable derivative of aldehydes or ketones for purposes of identification. All the other factors contribute to the choice and the key is **A**.

22. The reagent which reacts with an alcohol but not with an aldehyde is sodium, key **D**. Two of the substances in the list will react with both alcohols and aldehydes. Which are these two substances and what is formed in each case?

23. The reaction being tested here is the equivalent of hydrolysis of an ester, using ammonia (HNH_2) rather than the isoelectronic molecule water (HOH).

The correct key is **A**. The closest incorrect answer is **E**. What will be formed in this case?

24. The vigorous reaction with water, the formation of an acidic solution and the immediate release of halide ions, as shown by the silver nitrate test, are all reactions of an acyl halide. The correct key is **B**. None of the other substances will release bromide ions rapidly, hydrolysis of the aromatic halogen in **C** and **D** being particularly slow. None of the other substances will react vigorously with water and with only one of these four substances will a reasonably acidic solution appear. Which is the substance, apart from **B**, which gives an acidic solution?

25. The lack of violent reaction with water rules out the acyl chlorides, **A** and **B**. Failure to react readily with sodium carbonate rules out the acids **D** and **E**. The halogen atom in **C** is very difficult to hydrolyse. The phenolic —OH group in **C** is only weakly acidic. It is not sufficiently acidic to react with sodium carbonate. **C** is the correct key.

26. 0.01 mole of the compound reacts with 0.02 mole of NaOH. This shows that the molecule must contain, or be capable of producing, two acidic groups. This is only the case with **C**, **D** and **E**. The reaction with bromine water rules out **D** which does not contain a C=C double bond. The molecule must be **C** rather than **E** because of the loss of water on heating. This can take place readily between $-CO_2H$ groups if they are on the same side of the molecule. Is the arrangement in **C** called 'cis' or is it called 'trans'?

27. The formation of an oxime shows that the volatile decomposition product contains a carbonyl group. This will be propanone which is formed from calcium propanoate, key **D**. Can you write equations for the thermal decomposition of calcium propanoate and for the formation of the oxime?

28. Treatment of an alkyl halide with a sodium alkoxide is a general method for making ethers such as that given in key **C**. What is the other product of this reaction?

29. Heating ethanol with concentrated sulphuric acid under very similar conditions to that in the question can be used to prepare diethyl ether. The concentrated sulphuric acid is used in excess to minimize this second reaction. The key is **B**.

30. Another unwanted product of the reaction is SO_2, formed by a redox reaction between ethanol and concentrated sulphuric acid. The aqueous sodium hydroxide is to absorb this SO_2, key **D**.

31. The addition reaction between bromine and ethene is exothermic. Bromine is very volatile and it is necessary to cool the reaction mixture as much as possible to minimize loss of bromine by evaporation. The key is **E**.

32. The darkening of some organic reaction mixtures is due to liberation of traces of carbon. Key **C** comes closest to indicating the presence of carbon and this is the best choice.

33. Bromine will be more soluble in the organic layer of 1,2-dibromoethane than in the aqueous layer of **D** and **E**. Concentrated sulphuric acid, key **C**, will not remove bromine. What is required is a reagent that reacts with bromine molecules but does not react with 1,2-dibromoethane. The sodium hydroxide of **A** and **B** will react with both. In the absence of the ideal reagent, **A** will have to be selected as the best choice because of its slow reaction with the organic compound. Can you suggest the ideal reagent for removing bromine?

34. The bromine atoms in this compound are covalently bonded to carbon and are difficult to remove. Hydrolysis, assisted by alkali and warming, is suitable for releasing the bromine as bromide ions. Key **E** is the best choice. The use of acid in the silver nitrate solution prevents reaction of hydroxide ions with silver ions, allowing the reaction of bromide ions with silver ions to give the cream precipitate of silver bromide. Does it matter with which acid the silver nitrate is acidified? What would happen if the acid in the silver nitrate was insufficient to completely neutralize the sodium hydroxide?

35. Decomposition of the diazonium ion, key **E**, is the reason for working at low temperature. What is formed if the decomposition is allowed to take place?

36. Sodium thiosulphate will react with any iodine (key **A**) which may be present in small amounts due either to thermal decomposition of the iodobenzene or oxidation by air of iodine (-1) to iodine molecules.

37. 0·1 mole of aniline was used. The equations show that this could form 0·1 mole of iodobenzene, which is 20·4 g. A yield of 16·32 g is a percentage yield of $(16 \cdot 32/20 \cdot 4) \times 100$. This is close to 80 per cent, key **D**.

38. Many aromatic compounds can be made by using a diazonium salt at one stage in their preparation. A compound *usually* prepared from a diazonium salt is an azo compound. Benzene-azo-2-naphthol is an azo compound and the key is **C**.

Test 15

Carbon chemistry 3 (more difficult questions)

Questions 1–4 concern substances with the following formulae:

A $CH_3CH=CHCHO$

B $CH_3CH_2CO_2CH_2CH_3$

C $CH_3CHOHCO_2H$

D $CH_3COCH_2CO_2H$

E $CH_2=CHCO_2H$

Select the formula which most closely fits the description of each compound given below.

1. A colourless liquid which initially forms an immiscible layer with sodium hydroxide solution but, after refluxing for several hours, two organic products, a solid and a liquid, can be separated.

2. An optically active substance.

3. A colourless liquid which decolorizes acidified potassium manganate(VII) ($KMnO_4$) solution and forms a crystalline precipitate with 2,4-dinitrophenylhydrazine.

4. A liquid which reacts with metallic sodium evolving hydrogen and decolorizes both bromine water and acidified potassium manganate(VII) ($KMnO_4$) solution in the cold.

Questions 5–8

A substance may be classified as a

A monosaccharide

B polysaccharide

C fat-hydrolysing enzyme

D polysaccharide-hydrolysing enzyme

E protein-hydrolysing enzyme

Select from A to E, the heading under which you would classify each of the following:

5. carboxypeptidase

6. lysozyme

7. α-amylase

8. cellulose

Directions summarized for questions 9 to 11				
A	**B**	**C**	**D**	**E**
1,2,3 correct	1,2 only correct	2,3 only correct	1 only correct	3 only correct

9. Which of the following exhibit optical activity?

1   $CH_3-CH(NH_2)CH=CH-CH_3$

2   $CH_3-CH(Cl)CH(Cl)-C_2H_5$

3   $CH_3-C(Br)=C(Br)-CH_3$

10. Man-made polymers that are esters include

1   terylene

2   nylon

3   polystyrene

11. Carbonium ions are formed in the reactions

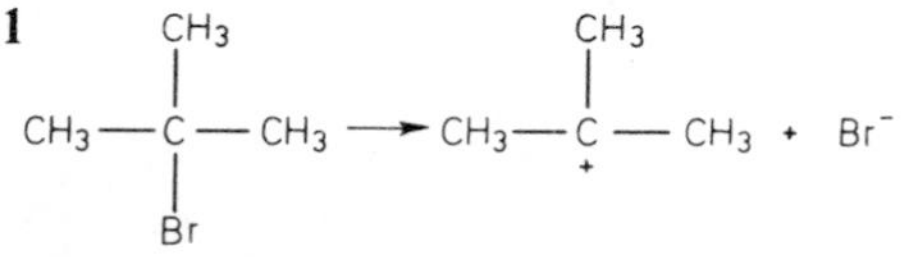

Directions summarized for questions 12 to 18				
A	**B**	**C**	**D**	**E**
1,2,3 only correct	1,3 only correct	2,4 only correct	4 only correct	Some other response or combination of responses is correct

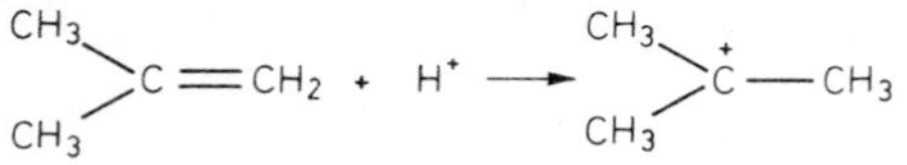

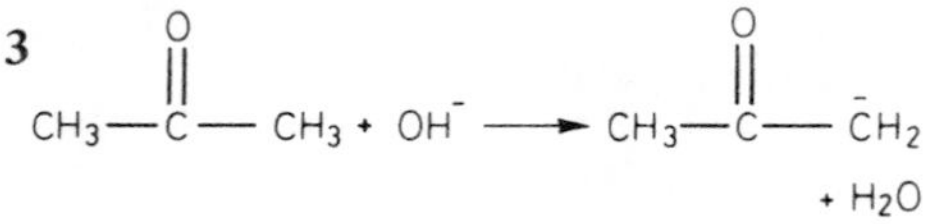

12. The mechanism of hydrolysis of the alkyl halide $(CH_3)_3CBr$ is thought to involve the following two steps:

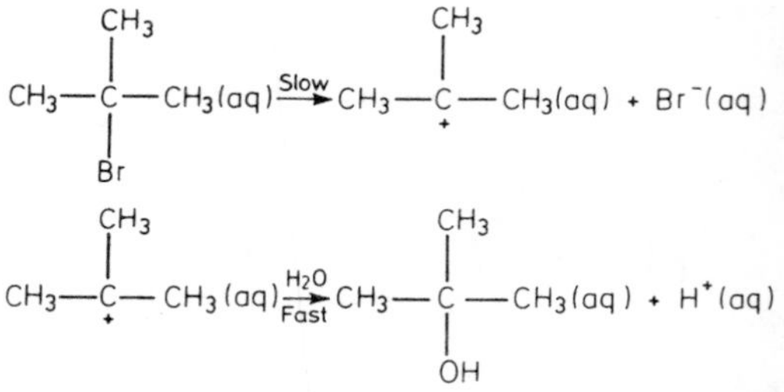

Which of the following statements are consistent with this mechanism?

1   The rate of hydrolysis is independent of the water concentration.

2   The rate of hydrolysis is directly proportional to the concentration of the alkyl halide.

3   Water is acting as a nucleophilic reagent.

4   The ion

$$CH_3-\underset{+}{C}-CH_3$$
$$|$$
$$CH_3$$

must be more stable than the alkyl halide but less stable than the alcohol.

13. Which of the following reactions would indicate that an acid X is maleic acid

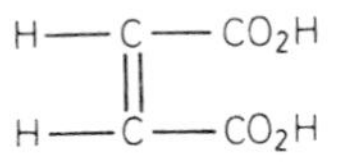

$$\begin{array}{l} \text{H} - \text{C} - \text{CO}_2\text{H} \\ \qquad \parallel \\ \text{H} - \text{C} - \text{CO}_2\text{H} \end{array}$$

and **NOT** fumaric acid

$$\begin{array}{l} \text{HO}_2\text{C} - \text{C} - \text{H} \\ \qquad\quad \parallel \\ \qquad \text{H} - \text{C} - \text{CO}_2\text{H ?} \end{array}$$

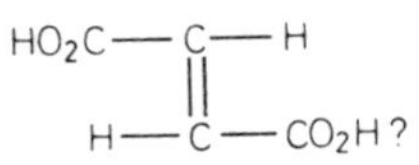

1 Reduction of X with sodium amalgam yields succinic acid,

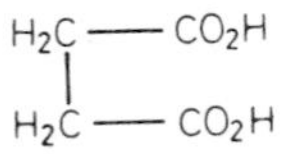

$$\begin{array}{l} \text{H}_2\text{C} - \text{CO}_2\text{H} \\ \quad | \\ \text{H}_2\text{C} - \text{CO}_2\text{H} \end{array}$$

2 Heating X with soda-lime yields ethene,

$$\text{H}_2\text{C}{=}\text{CH}_2$$

3 Reaction of X with hydrogen bromide yields bromosuccinic acid,

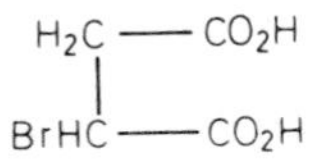

$$\begin{array}{l} \text{H}_2\text{C} - \text{CO}_2\text{H} \\ \quad | \\ \text{BrHC} - \text{CO}_2\text{H} \end{array}$$

4 Gentle heating of X under slightly reduced pressure yields the acid anhydride,

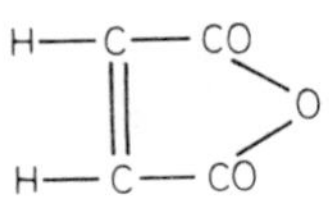

$$\begin{array}{l} \text{H} - \text{C} - \text{CO} \\ \qquad \parallel \qquad\quad \text{O} \\ \text{H} - \text{C} - \text{CO} \end{array}$$

14. Ozone is a form of oxygen in which the molecules are triatomic. Ozone reacts with compounds containing carbon-to-carbon double bonds thus:

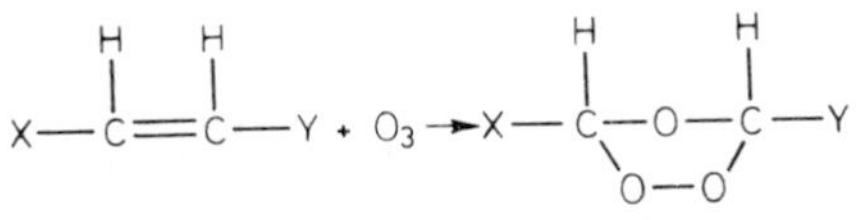

$$\text{X} - \underset{|}{\overset{|}{\text{C}}}{=}\underset{|}{\overset{|}{\text{C}}} - \text{Y} + \text{O}_3 \rightarrow \text{X} - \underset{\diagdown}{\overset{|}{\text{C}}}\underset{\text{O}-\text{O}}{-\text{O}-}\underset{\diagup}{\overset{|}{\text{C}}} - \text{Y}$$

The product of this reaction can be carefully hydrolysed as shown:

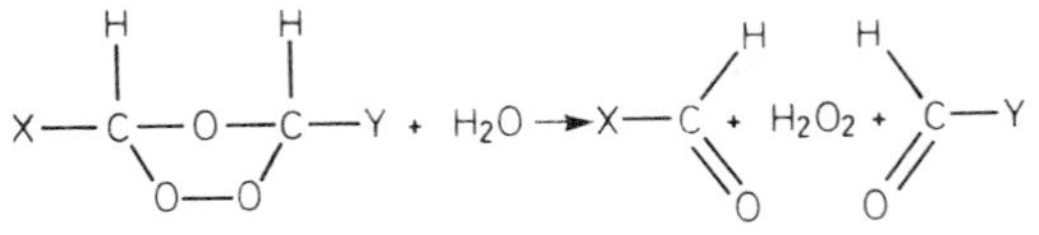

$$\text{X} - \underset{\diagdown}{\overset{|}{\text{C}}}\underset{\text{O}-\text{O}}{-\text{O}-}\underset{\diagup}{\overset{|}{\text{C}}} - \text{Y} + \text{H}_2\text{O} \rightarrow \text{X} - \overset{|}{\underset{\parallel}{\text{C}}} + \text{H}_2\text{O}_2 + \overset{|}{\underset{\parallel}{\text{C}}} - \text{Y}$$

Which of the following would be reasonable uses of either or both of these reactions?

1 To find out if a compound contains a carbonyl group

2 To find the percentage of ozone in a sample of partially ozonised oxygen

3 To promote polymerization of compounds containing double bonds

4 Determining the location of double bonds in suitable compounds

15. Correct statements about the nitration of benzene include that

1 attack on the benzene ring is by a nucleophilic reagent

2 the attacking entity is NO_2^-

3 the intermediate and the product contain the same delocalized electron system

4 a proton is eliminated from the intermediate

16. The compound of formula

$$\text{CH}_3\text{CO(CH}_2)_5\text{CH}{=}\text{CHCO}_2\text{H}$$

would be expected to

1 give tri-iodomethane (iodoform) with iodine and potassium hydroxide

2 decolorize bromine water

3 have *cis*- and *trans*-isomers

4 exist in optically active forms

17. The formula of an organic compound is

$$CH_3 CHCH_2 C(C_2 H_5)_2 CH_2 CO_2 H$$
$$\underset{NH_2}{|}$$

This compound

1 is capable of forming a polymer

2 decolorizes potassium permanganate solution

3 is capable of reaction BOTH with sodium hydroxide and with hydrochloric acid

4 gives a yellow precipitate with 2,4-dinitrophenylhydrazine

18. The presence of a protein or other polypeptide is shown by the formation of

1 a red precipitate with boiling Fehling's solution

2 a deep blue colour with iodine solution

3 a deeper blue colour when it is added to copper(II) sulphate solution

4 a purple colour with sodium hydroxide and a very little copper(II) sulphate solution

Directions for questions 19 to 28. Each of the questions or incomplete statements in this section is followed by five suggested answers. Select the best answer in each case.

19. Which is the order of INCREASING acid strength of the compounds below?

I $C_6 H_5 OH$

II $CH_3 CO_2 H$

III $C_2 H_5 OH$

IV $HCO_2 H$

V $Cl_3 CCO_2 H$

A I–III–IV–V–II

B II–IV–V–I–III

C III–I–II–IV–V

D IV–II–III–I–V

E V–I–III–IV–II

20. The reaction between aqueous sodium hydroxide and bromoethane is best described as

A a free radical reaction

B esterification

C nucleophilic attack by hydroxide ions

D nucleophilic attack by sodium ion

E electrophilic attack by hydroxide ions

21.

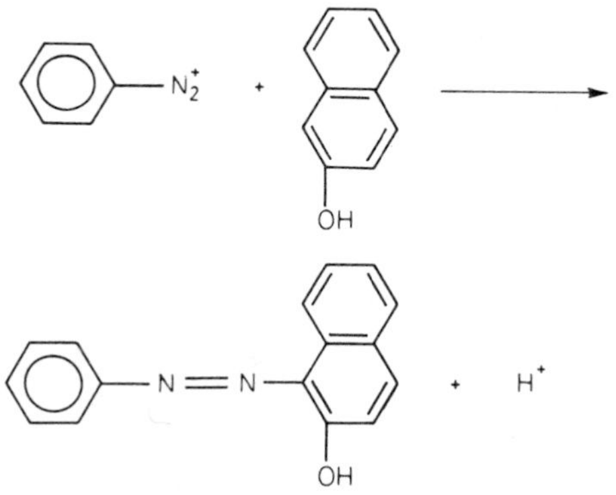

In the reaction represented by the equation, into which one of the following types of compound could the organic product be classified?

A A thermosetting plastic

B A dyestuff

C An explosive

D A starting material for making detergents

E An insecticide

22. An organic compound R is found to behave as follows:

I on oxidation of R, a compound was formed which gave a precipitate with 2,4-dinitrophenylhydrazine but did not reduce Fehling's solution.

II on treatment of R with nitrous acid at 5 °C followed by addition of 2-naphthol (2-hydroxynaphthalene) in sodium hydroxide solution, an orange precipitate was formed.

R could be

A

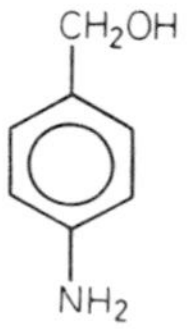

B

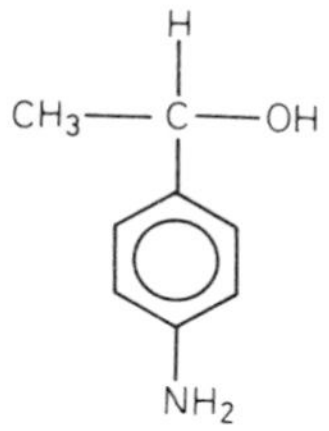

C

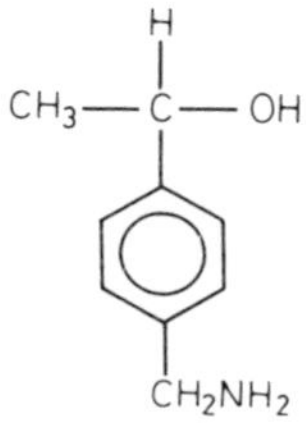

D

E

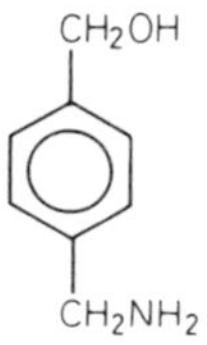

23. A substance has the following properties:

It reacts with sodium hydroxide to form an ionic solid.

It reacts with phosphorus pentachloride, giving off hydrogen chloride.

It reacts with hydrogen by addition in the presence of a nickel catalyst.

It reacts with ethanol.

The substance could be

A CH_2=CH—$\overset{\displaystyle O}{\overset{\|}{C}}$—$OH$

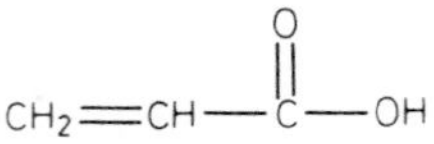

B (benzene ring)—OH

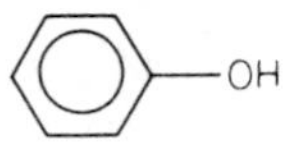

C CH_2=CH—CH_2—OH

D CH_3—$\overset{\displaystyle O}{\overset{\|}{C}}$—$H$

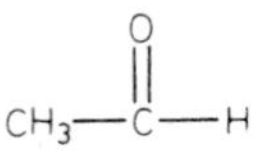

E CH_3—$\overset{\displaystyle O}{\overset{\|}{C}}$—$OH$

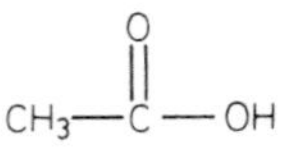

24. Suppose the following five substances were under consideration for the preparation of a high molecular weight polymer:

I CH_3—$\overset{\displaystyle }{\underset{\displaystyle \underset{OH}{\underset{|}{CH_2}}}{\overset{|}{C}}}$=$CH_2$

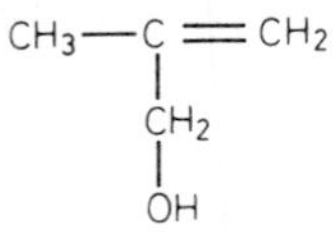

II Cl—CO—CH_2—CH_2—CO—Cl

III H_2N—CH_2—(benzene ring)—CH_2—NH_2

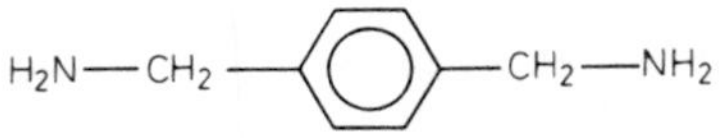

IV HO—CH_2—$CHOH$—CH_2—OH

V H_2N—CH_2—$\underset{\displaystyle \underset{CH_3}{\underset{|}{}}}{\overset{|}{CH}}$—$CO$—$Cl$

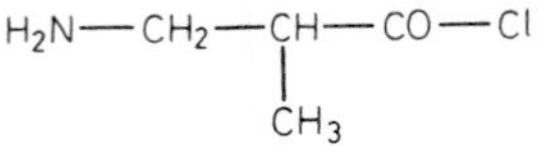

In which of the following cases is a high molecular weight polymer LEAST likely to be formed?

A I alone

B II and IV reacting together

C II and III reacting together

D II alone

E V alone

25. An organic compound has the following properties

I   it reacts with phosphorus pentachloride to form hydrogen chloride.

II   it decolorizes acidified potassium permanganate.

III   it is optically active.

The organic compound could have the formula

A $CH_3CH(OH)CO_2H$

B $(CO_2H)CH$=$CH(CO_2H)$

C $CH_2NH_2CO_2H$

D $(CO_2H)_2$

E $CH_3CH_2CH_2CH_2OH$

26. An organic compound has the following properties:

I   it decolorizes a solution of bromine in tetrachloromethane.

II   it reacts with ethanolic potassium hydroxide solution to form a compound with two C=C double bonds.

III   it is hydrolysed, with boiling water or an alkali, to a compound with two alcoholic —OH groups.

The structure of the organic compound could be

A $ClCH_2CH_2CH=CHCO_2H$

B $ClCH_2CH_2CH=CHCH_2OH$

C $BrCH_2CH_2CH_2CH_2COCl$

D $HOCH_2CH=CHCH_2CH_2OH$

E $CH_2=CHCH=CHCHCl_2$

27. The reason why phenylamine (aniline) is a much weaker base than ammonia when each is in aqueous solution is that

A   the lone-pair of electrons on the nitrogen atom of phenylamine is delocalized over the benzene ring

B   acids substitute into the benzene ring of phenylamine rather than form salts

C   the phenylamine molecule is too large to capture hydrogen ions easily

D   phenylamine is much less soluble in water than is ammonia

E   the benzene ring has a tendency to increase the acidity of its substituents

28. The substance of a formula

$$(...CH_2CH_2O_2CC_6H_4CO_2CH_2CH_2O_2CC_6H_4CO_2...)_n$$

is a

A   polyester   D   detergent

B   rubber   E   protein

C   natural oil or fat

29. A mixture of two amino acids

$$NH_2-CH-CO_2H \quad NH_2-CH-CO_2H$$
$$\quad\quad\quad | \quad\quad\quad\quad\quad\quad\quad\quad |$$
$$\quad\quad (CH_2)_3 \quad and \quad\quad (CH_2)_3$$
$$\quad\quad\quad | \quad\quad\quad\quad\quad\quad\quad\quad |$$
$$\quad\quad\quad NH_2 \quad\quad\quad\quad\quad\quad CO_2H$$
$$\quad\quad\quad (I) \quad\quad\quad\quad\quad\quad\quad (II)$$

is dissolved in an excess of 2 M hydrochloric acid and subjected to paper electrophoresis. Which of the following best describes what happens?

A   Both move at the same rate towards the cathode.

B   II moves towards the anode faster than I.

C   I moves towards the cathode faster than II.

D   Both move at the same rate towards the anode.

E   II moves towards the cathode faster than I.

30. An amino-acid NOT occurring in natural proteins is

A $H_2N-CH-CO_2H$
$\quad\quad\quad\quad |$
$\quad\quad\quad\quad H$

B $H_2N-CH-CO_2H$
$\quad\quad\quad\quad |$
$\quad\quad\quad\quad CH_3$

C $H_2N-CH_2-CH-CO_2H$
$\quad\quad\quad\quad\quad\quad\quad |$
$\quad\quad\quad\quad\quad\quad CH_2-NH_2$

D $H_2N-CH-CO_2H$
$\quad\quad\quad\quad |$
$\quad\quad\quad\quad CH_2OH$

E $H_2N-CH-CO_2H$
$\quad\quad\quad\quad |$
$\quad\quad\quad\quad CH_2SH$

31. For which of the following compounds is the rate of hydrolysis by aqueous alkali most likely to be independent of the hydroxide ion concentration?

    A 1-chlorobutane

    B 1-bromobutane

    C 1-iodobutane

    D 2-chlorobutane

    E 2-bromo-2-methylbutane

32. Which one of the following methods could NOT be used to determine the rate of hydrolysis of a simple alkly iodide, RI, by aqueous alkali?

    A Measurement of the electrical conductivity of the reacting mixture

    B Measurement of the colour of the reacting mixture

    C Measurement of the pH of the reacting mixture

    D Titration of samples of the reacting mixture with standard acid

    E Titration of samples of the reacting mixture (after neutralization) with standard, aqueous silver nitrate

Questions 33–35 concern a comparison of the rates of hydrolysis of the halogenoalkanes, 1-chloro-butane, 1-bromobutane and 1-iodobutane. The following apparatus was used:

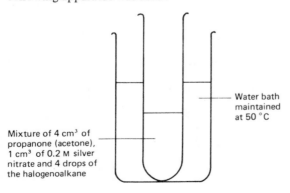

Mixture of 4 cm³ of propanone (acetone), 1 cm³ of 0.2 M silver nitrate and 4 drops of the halogenoalkane

Water bath maintained at 50 °C

The beaker was placed on a square of white paper on which was marked a black cross. The time for the contents of the test tube to mask the black cross was measured.
The experiment was repeated for each halogenoalkane and the results were as follows:

1-chlorobutane	15 minutes
1-bromobutane	2·5 minutes
1-iodobutane	3 seconds

33. Which of the following represents the best order in which the various operations should be carried out?

    A Warm all reagents to 50 °C; add 0·2 M silver nitrate to the test tube, followed by the halogenoalkane and then the propanone; start the clock.

    B Add the 0·2 M silver nitrate to the test tube followed by the halogenoalkane; allow to warm to 50 °C and add the propanone; start the clock.

    C Warm all the reagents to 50 °C; add the propanone to the test tube, followed by the halogenoalkane and then the 0·2 M silver nitrate; start the clock.

    D Add the halogenoalkane to the test tube followed by the 0·2 M silver nitrate; allow to warm to 50 °C and add the propanone; start the clock.

    E Warm all reagents to 50 °C; add the halogenoalkane to the test tube, followed by the 0·2 M silver nitrate and then the propanone; start the clock.

34. Which of the following is the best explanation of the observed differences in reaction rate?

    A The three halogen atoms differ in radius.

    B Different numbers of moles of the three halogenoalkanes were used.

    C The three reactions take place by different mechanisms.

    D The three halogenoalkanes differ in the extent of their solubility in water.

    E Bond strengths decrease in the order C–Cl, C–Br and C–I.

35. The hydrolysis of the halogenoalkane is actually brought about by

A propanone molecules

B silver ions

C hydrogen ions

D hydroxide ions

E nitrate ions

Questions 36—39 are concerned with the article below. You are advised to look at the questions before scanning the article.

NATURAL NITROGEN-FIXATION

Thanks to a happy accident, biological nitrogen-fixation may soon be understood. Years of research have established that some bacteria contain ENZYMES that reduce atmospheric nitrogen to ammonia and biological molecules. The details are not yet clear, but it is known that there are two main components of the enzyme system, one ▨ 1 ▨ containing iron and molybdenum and one just iron.

R. C. Burns and colleagues have recently prepared large quantities of the enzyme nitrogenase from *Azobacter vinelandii*, separating from it a crude fraction of the molybdenum-iron catalyst. And then, as they relate, when they lowered the salt concentration, the ▨ 1 ▨ crystallized out in large regular crystals.

This makes all sorts of new studies possible. Most interesting is the possibility of determining the enzyme's three-dimensional structure by ⬭ 2 ⬭. They might then discover how nature fixed nitrogen at room temperature and pressure, whereas the chemical industry needs high pressures and temperatures. The world market for ⬡ 3 ⬡ is enormous, and the company that first discovers how biological molybdenum–iron catalysts work, and makes its own version, will gain a huge profit.

36. Which of the following words would best fit in each space marked ▨ 1 ▨ ?

A Peptide

B Hydrocarbon

C Amino-acid

D Protein

E Carbohydrate

37. Which of the following techniques is likely to be referred to in the space marked ⬭ 2 ⬭

A *X*-ray diffraction

B Infrared spectrometry

C Hydrolysis followed by two-dimensional paper-chromatography

D Mass spectrometry

E Electron microscopy

38. Which of the following fits best in the space marked ⬡ 3 ⬡ ?

A Molybdenum

B Enzymes

C Hormones

D Complex iron compounds

E Nitrogenous fertilizers

39. Which of the following is most closely related in structure and biochemical function to the type of substance discussed in the article?

A 'DNA'

B 'LSD'

C Insulin

D Penicillin

E Nicotine

Test 15   Answers

1. All five compounds react with aqueous sodium hydroxide. **C, D** and **E** would be expected to dissolve rapidly, forming sodium salts. Choice has to be made between **A** and **B**. **A** contains a hydrogen atom on the carbon atom next to the aldehyde group and would be expected to polymerize rather than taking part in the Cannizzaro reaction. The correct key is **B**. The ester is slowly hydrolysed to ethanol, which could be distilled out as a liquid, and sodium propanoate, which could be crystallized out as a solid.

   If **A** had been involved in the Cannizzaro reaction, would this compound also have fitted the description?

2. Optical activity can occur when a structure has no symmetry. In carbon chemistry this is most frequently met when a saturated carbon atom is attached to four different groups. This is the case in **C** which is the correct key.

   One of the molecules would be expected to be able to exist in the form of a *cis-* and a *trans-*isomer. Which molecule is this and why is a rather similar molecule in the list not able to show this type of isomerism?

3. The decolorization of acidified permanganate suggests the presence of an alkene, alkyne, primary alcohol, secondary alcohol or aldehyde (**A, C** or **E**). The reaction with 2,4-dinitrophenylhydrazine shows that an aldehyde (**A**) or ketone (**D**) is present. **A** is an alkene and an aldehyde and must be the correct key.

4. The reaction with metallic sodium indicates an acid or alcohol (**C, D** or **E**). The bromine water reaction suggest an alkene or alkyne (**A** or **E**), while the permanganate reaction suggests, as in question **3**, that the molecule is **A, C** or **E**. The correct key is **E** which is suggested by all three tests.

5. Carboxypeptidase is an enzyme which hydrolyses peptide links, working steadily through a protein chain from the free carboxyl end. The release of amino acids by the enzyme gives

information about their sequence in the protein. The key is **E**.

6. Lysozyme is a polysaccharide-hydrolysing enzyme, key **D**. It attacks certain bacteria by breaking down the polysaccharide in their cell walls. The 129 amino acid sequence in lysozyme is almost constant over a wide variety of species. Slight differences in the sequence give information about the time that has passed since species diverged from common ancestors.

7. α-Amylase is an enzyme that hydrolyses unbranched polysaccharide chains in starch, key **D**, breaking down the molecule to maltose. This enzyme is present in saliva.

8. Cellulose is a polysaccharide, key **B**.

9. Optical activity is possible when a molecule is completely lacking in symmetry. **1** and **2** fit this description as each contains at least one carbon atom attached to four different groups. The key is **B**.

10. Terylene is the only polymer in the question that is an ester. Nylon is an amide and polystyrene is a hydrocarbon. The key is **D**.

11. Carbonium ions are the positive ions of carbon, believed to exist as short-lived species at very low concentration during many organic reactions. **1** and **2** show the formation of carbonium ions. The key is **B**.

12. This is a reaction in which carbonium ions are believed to be involved. The slow step would determine the rate of hydrolysis. The rate equation that is consistent with this step is

$$\text{Rate} = k[\text{water}]^0[\text{alkyl halide}]^1$$

   This equation shows that **1** and **2** are correct. **3** is also correct — in the fast step, a lone pair on the oxygen atom of the water molecule is assisting this nucleophilic reagent in its attack on the centre of positive charge. What is a reagent called if it attacks a part of a molecule or ion where electron pairs are available?

It is not possible to make a statement about the relative stabilities of the species involved in the reaction. To what extent would **4** be correct and consistent if the word 'stable' was replaced by 'reactive' both times?

1, 2 and 3 are correct and the key is **A**.

13. Only reaction 4 depends on the two carboxyl groups being on the same side of the molecule in the *cis*-form, maleic acid. The key is **D**. Loss of water takes place far more readily in this molecule compared with the *trans*-form, where forcing conditions would be required to persuade rotation about the double bond to take place. This rotation, which would give the *cis*-form, is necessary before loss of water to give the anhydride can occur.

14. Only 2 and 4 are correct, key **C**. The percentage of ozone in partially ozonized oxygen could be found by first taking a known volume of the gaseous mixture. This would be allowed to react with the alkene. The quantity of alkene consumed, the decrease in volume as the ozone reacts, the quantity of aldehyde formed or the quantity of hydrogen peroxide produced could all be used to determine how much ozone was present. Which of these four measurements would you make if you had a choice?

The location of the double bonds could be inferred from the exact nature of the aldehydes XCHO and YCHO formed on hydrolysis.

15. During nitration, benzene is believed to be attacked by the electrophilic reagent, NO_2^+ (1 and 2 both incorrect). It is as though the NO_2^+ persuades the delocalized ring system to make a lone pair available momentarily in the vicinity of the NO_2^+.

NO₂⁺ approaches delocalized ring system

Delocalized system disappears. Electron pairs located in specific positions

The intermediate. Electron pairs located in specific positions

The product. Delocalized system returns

In the mechanism suggested here, the intermediate has lost the delocalized ring system (3 incorrect). This system is regained by loss of a proton as the intermediate turns to the final product (4 correct). The key is **D**.

The attack of the benzene ring by the NO_2^+ ion can be regarded as a reaction between a Lewis acid and a Lewis base. Which substance is the acid and which is the base?

16. The presence of the CH_3COR group means that the iodoform reaction is to be expected (1 correct). The C=C double bond should decolorize bromine water (2 correct). The structure of the type XCH=CHY would allow *cis*- and *trans*-isomers to exist (3 correct). There is no centre of asymmetry and **4** is incorrect so the key is **A**.

17. The molecule does not contain a C=C double bond or any other group that would be expected readily to reduce permanganate, and 2 is incorrect. The carbonyl group of aldehydes or ketones is not present which rules out a reaction with 2,4-dinitrophenylhydrazine and 4 is incorrect.

1 and 3 are correct (key **B**). The molecule has two reactive functional groups, $-CO_2H$ and $-NH_2$. These are capable of reacting with each other to allow polymerization and with NaOH and HCl to form salts.

18. 1 is a reaction of aldehydes, 2 is a test for starch and 3 is a reaction of ammonia and amines when they form complexes with $Cu^{2+}(aq)$.

Only 4 is correct (key **D**). The reaction is known as the biuret test and depends on the formation of a complex ion between $Cu^{2+}(aq)$ and the protein, with the peptide groups acting as ligands.

19. Trichloroethanoic acid is by far the strongest of the acids and **V** must come last. This fact alone places the correct key as **C** or **D**.

Ethanol and phenol must come first, in that order, as they are by far the weakest. Ethanoic acid and methanoic acid must come next and you do not need to know which way round these come if you have used the other information. The key is **C**.

20. The alkaline hydrolysis of bromoethane involves loss of Br^-, leaving a carbon atom with a momentary positive charge. Depending on the mechanism, S_N1 or S_N2, simultaneous or subsequent attack by OH^- ions takes place. This is nucleophilic attack (key **C**) because the OH^- ions are seeking an electron deficient centre. What do the letters and numbers stand for in the mechanisms S_N1 and S_N2 mentioned above?

21. The reaction is forming an azo-dye, key **B**. Is the diazonium ion acting as a nucleophilic or as an electrophilic reagent in this reaction?

22. Test **I** shows that R can be oxidized to a ketone rather than an aldehyde. R must be a secondary alcohol and will be **B** or **C**.

    Test **II** shows that R will form a diazonium ion which means that it must be a primary aromatic amine. R will be **A**, **B** or **D**.

    The only solution that fits both tests is key **B**.

23. The carboxylic acids **A** and **E** will react with NaOH to form ionic salts, as will the slightly acidic phenol, **B**, and possibly the alcohol, **C**. Alcohols react reversibly with NaOH to form strongly hydrolysed salts, in this case $CH_2{=}CH{-}CH_2{-}O^-Na^+$. **D** would polymerize rather than form a salt.

    PCl_5 would certainly produce HCl with the carboxylic acids and the alcohol, **A**, **C** and **E**. Phenol gives some HCl. **D** reacts but does not produce HCl.

    Hydrogen and a nickel catalyst will reduce the C=C double bond in **A** and **C**. This will also reduce the aromatic ring in **B** and the aldehyde **D**.

    Under suitable conditions, the acids will react with ethanol to form esters and the aldehyde can undergo addition on the C=O group to form an acetal.

    The only substance which reacts, as described in the question, in all four tests is **A**. This is the correct key.

24. An organic molecule has the potential to become involved in polymer formation if it possesses two reactive functional groups. These two groups include the two 'ends' of a double bond, and I alone (**A**) can polymerize. The addition polymerization of I is similar to the conversion of ethene to polyethene.

    The possession of two reactive functional groups only gives the potential to polymerize. This potential can be realized if the functional groups are capable of a sustained reaction which allows polymer chains to grow. This growth is possible with II and IV (**B**). An ester link is formed between this pair while, at the same time, $-COCl$ and $-OH$ groups remain at the ends of the growing molecule and available for further reaction. A similar situation arises with II and III (**C**) where the amide link, $-CO-NH-$, is formed. This link is also formed when V alone polymerizes (**E**).

    The molecule in II contains two reactive groups, both $-COCl$, but the formation of a new covalent bond is not possible between a pair of these groups. II is unlikely to polymerize and the key is **D**.

25. The first two properties are not particularly helpful! If the conditions are right, all five molecules could form HCl by reaction with PCl_5 as they all contain an $-OH$ group. (Which of these molecules might have lost the $-OH$ group by formation of an internal ion?) The decolorization of acidified permanganate is shown by reducing agents, which include a secondary alcohol (**A**), the C=C double bond, (**B**), ethanedioic acid (oxalic acid) (**D**) and a primary alcohol (**E**).

    The answer to the question is decided by III. Only **A** is capable of showing optical activity as it alone possesses the structure of four different groups attached to a carbon atom.

26. I shows that the molecule can be **A**, **B**, **D** or **E**. The reaction with alcoholic KOH will eliminate halogen acid from adjacent C–halogen and C–H groups, forming an extra double bond. Two double bonds will be present after this reaction in **A** and **B**. (The $-COCl$ group in **C** will not take part in loss of HCl to form a double bond, but will hydrolyse to $-CO_2^-K^+$. **E** is likely to become a molecule with three C=C double bonds.)

Properties I and II show that the molecule must be **A** or **B** but only **B** will be hydrolysed to a compound with two alcoholic —OH groups, one present before reaction, the other formed by hydrolysis.

As in question 25, the answer can be decided on the final piece of evidence alone. **B** and **D** are the only two molecules that would contain two —OH groups after hydrolysis had been attempted but **D** could not be described as 'hydrolysed' as both —OH groups are present before hydrolysis. If you were going to answer the question using III alone, you would need to know what is formed on hydrolysis of the group —$CHCl_2$. The first product of hydrolysis could be regarded as —$CH(OH)_2$ but this contains two —OH groups on the same carbon atom. Compounds with this structure are usually unstable and when attempts are made to isolate them by distillation, loss of water occurs and the product isolated is the aldehyde, —CHO.

27. The present understanding of the situation would favour key **A**. The lone pair of electrons on the nitrogen atom is believed to be less available for donation because it is to some extent involved in the bonding of the benzene ring. Three of the structures below show the lone pair at different positions in the ring. The lone-pair would be described as 'delocalized over the benzene ring'.

28. The substance contains the repeating link —O—CO— and is a polyester, key **A**.

29. The treatment with hydrochloric acid will convert the —NH_2 groups of the amino acids to —NH_3^+ groups.

$$-NH_2 + H_3O^+ \rightleftharpoons -NH_3^+ + H_2O$$

Both molecules will become positively charged and will therefore move towards the negative cathode during electrophoresis. B and D are incorrect.

Molecule I contains two —NH_2 groups and will pick up a double positive charge on treatment with hydrochloric acid. This makes it likely that I will move faster towards the cathode than II, which will have picked up only one positive charge on acid treatment. The correct key is **C**.

30. In all naturally occurring amino acids, the —NH_2 group is attached to the carbon atom adjacent to the —CO_2H group. This is not the case in **C** which is the correct key.

Which one of these amino acids is likely to move fastest towards the cathode during electrophoresis after treatment with hydrochloric acid? Can you identify which three of these amino acids are able to show optical activity?

31. Alkaline hydrolysis of alkyl halides involves nucleophilic substitution. The OH^- ion acts as a nucleophile, seeking out the partial positive charge that is found on the carbon atom to which the halogen atom is attached. This is the slow step in the reaction and the rate equation is of the form

$$\text{Rate} = k \, [OH^-]^1 \, [\text{alkyl halide}]^1$$

In such a reaction, the rate of hydrolysis is dependent on the hydroxide concentration.

With some alkyl halides, however, the carbon atom to which the halogen is attached appears to acquire a complete positive charge at some stage in the reaction. The alkyl halide molecule becomes a carbonium ion by loss of the negative halide ion as shown in the first equation of question 11 and 12 on page 172. This is the slow step in such a hydrolysis. The later stage, when a hydroxide ion or a water molecule attacks the carbonium ion, is fast. The rate equation is then of the form.

$$\text{Rate} = k \, [OH^-]^0 \, [\text{alkyl halide}]^1$$

In such a reaction, the rate of hydrolysis is independent of the hydroxide ion concentration.

The second mechanism is most likely to occur with alkyl halides containing three bulky groups attached to the carbon atom to which the halogen atom is also attached. This is because formation of the planar carbonium ion relieves some of the strain of repulsion between the bulky groups that was present in the tetrahedral arrangement of the original molecule.

Tetrahedral arrangement around central carbon atom with bond angles of about 109°.

Planar arrangement around central carbon atom with bond angles of about 120°. Reduced repulsion between alkyl groups.

Of the alkyl halides in the question, 2-bromo-2-methylbutane is the one where the repulsional influence between alkyl groups is likely to be greatest. It is the only molecule in the list which has three alkyl groups attached to the central carbon atom. This is the halide that is most likely to form a carbonium ion and hence hydrolyse by a mechanism whose rate is independent of the hydroxide ion concentration. The correct key is **E**.

The two steps in a hydrolysis involving carbonium ions are shown in question 12. Is the mechanism of this reaction S_N1 or is it S_N2?

32. Rates of reaction can be investigated by measurement of the rate at which one of the reactants disappears or the rate at which one of the products appears. In the alkaline hydrolysis of an alkyl iodide, the rate could be determined by finding either the rate of disappearance of OH^- ions (**C** and **D**) or the rate of appearance of I^- ions (**E**). Measurement of electrical conductivity could also be used to find how rapidly the highly conducting OH^- ion is being replaced by the I^- ion of far lower conductivity (**A**).

Simple alkyl iodides do not have a colour and the organic reaction product, an alcohol, is also colourless. Method B cannot be used to follow the rate of this reaction and the key is **B**.

You have probably used an alkyl iodide when investigating the relative rates of hydrolysis of alkyl chlorides, bromides and iodides. If so, it is likely that the alkyl iodide had a brown colour. This is the result of slight oxidation by the air which forms iodine. What tests would you carry out on the brown alkyl iodide to see whether the pure compound was colourless and whether the brown colour was due to iodine?

33. The purpose of the propanone is to act as a solvent for both the organic and aqueous reagents. One reagent should be dissolved in the propanone and the other reagent then added. Only in **C** is this done.

Both reagents should be allowed to reach the experimental temperature separately before allowing to react. **A**, **C** and **E** are correct here.

The clock should be started as soon as the aqueous layer has been added to the halogenoalkane as this is the point when hydrolysis starts. This only happens in **C**. **C** must be the correct key.

There is a weak point about the experiment as the reaction carried out in this way may not be simple hydrolysis. The presence of silver ions may alter the mechanism of the reaction and the rate.

34. **A** is true but makes no attempt to explain the observed differences. **B** is also true but the chloroalkane will have the smallest molar volume so this substance will have been taken in the largest number of moles. Despite this, the rate of hydrolysis is the slowest. The rate of hydrolysis is fastest for the iodoalkane although it is being taken in the smallest number of moles. **B** does not explain the observations.

C might be true. There are, however, only two mechanisms between which to choose, S_N1 and S_N2. The three reactions could not all take place by different mechanisms unless, perhaps, one was S_N1, another S_N2 and the third a mixture of the two mechanisms. There is the possibility of a free radical mechanism but, even if three mechanisms were available, **C** makes no attempt to explain the observed differences.

D cannot be considered as the propanone is present to avoid any solubility problems.

E is the correct key. The statement about bond strengths is correct and this does offer an explanation for the observed differences in reaction rate.

35. It is possible, as mentioned in the answer to question 33, that the silver ions interfere with the normal course of the hydrolysis. But of the particles between which choice has to be made, hydroxide ions, key **D**, would be regarded as the best choice.

There is another nucleophilic reagent present in the reaction. This is water. The concentration of water molecules is about 10^9 times greater than the concentration of hydroxide ions in this experiment. Perhaps hydroxide ions do not really bring about the hydrolysis. This does not alter the choice of **D** as the correct key.

36. 1 will be a protein, key **D**.

37. The technique for finding the three-dimensional structure of a solid involves *X*-ray diffraction, key **A**.

38. The market referred to is that for nitrogenous fertilizers, key **E**.

39. Insulin, key **C**, is the substance most closely related to the enzymes discussed in the article.